The manufacturer's authorised representative in the EU is Springer Nature Customer Service Centre GmbH, Europaplatz 3, 69115 Heidelberg, Germany. If you have any concerns regarding our products, please contact ProductSafety@springernature.com

Printed and bound by CPI Group (UK) Ltd, Croydon, CR0 4YY
30/09/2026
02225607-0006

Green Energy and Technology

Climate change, environmental impact and the limited natural resources urge scientific research and novel technical solutions. The monograph series Green Energy and Technology serves as a publishing platform for scientific and technological approaches to "green"—i.e. environmentally friendly and sustainable—technologies. While a focus lies on energy and power supply, it also covers "green" solutions in industrial engineering and engineering design. Green Energy and Technology addresses researchers, advanced students, technical consultants as well as decision makers in industries and politics. Hence, the level of presentation spans from instructional to highly technical.

Indexed in Scopus.

More information about this series at http://www.springer.com/series/8059

Inamuddin · Rajender Boddula ·
Abdullah M. Asiri
Editors

Sustainable Ammonia Production

Springer

Editors
Inamuddin
Department of Chemistry
Faculty of Science
King Abdulaziz University
Jeddah, Saudi Arabia

Department of Applied Chemistry
Faculty of Engineering and Technology
Aligarh Muslim University
Aligarh, India

Abdullah M. Asiri
Department of Chemistry
Faculty of Science
King Abdulaziz University
Jeddah, Saudi Arabia

Rajender Boddula
CAS Key Laboratory of Nanosystem
and Hierarchical Fabrication
Beijing, China

ISSN 1865-3529 ISSN 1865-3537 (electronic)
Green Energy and Technology
ISBN 978-3-030-35105-2 ISBN 978-3-030-35106-9 (eBook)
https://doi.org/10.1007/978-3-030-35106-9

This Springer imprint is published by the registered company Springer Nature Switzerland AG
The registered company address is: Gewerbestrasse 11, 6330 Cham, Switzerland

Contents

Earth Abundant Catalysis for Ammonia Synthesis

Bilge Coşkuner Filiz and Aysel Kantürk Figen

Abstract The sustainable ammonia (NH_3) synthesis is not only one of the most attractive processes but also one of the most challenging catalytic ones under ambient conditions. The exothermic characteristic of synthesis reaction and also stability and inert behaviour of atmospheric nitrogen (N_2) make the conversion of N_2 to NH_3 hard, while N_2 is available in 78% in air. The industrial operations have been conducted under high temperature–pressure conditions by conventional Haber–Bosch process. The high energy requirement due to harsh operating conditions and the evolution of greenhouse gases (e.g. CO_2) during the synthesis make this process unsustainable for NH_3 synthesis. Besides, this process has made a lot of contribution to the catalysts field for nourishing, the sustainable and novel improvements have been still looked for more ambient and green synthesis process. The low synthesis efficiency and harsh operating conditions depend on the process that has still required to be improved and attracted many researchers' interests. In this chapter, earth abundant catalysis for NH_3 synthesis was gotten the point of classical and sustainable process approaches.

Keywords Earth abundant metal · Catalyst · Ammonia · Synthesis

1 Introduction

Ammonia (NH_3) synthesis has been grown one of the most major processes that have been used more than 1% of power consumption all over the world. Since 1910, the studies on the catalytic hydrogen and nitrogen reaction for the synthesis of NH_3 have been still gone on. Up to today, catalytic ammonia synthesis has led to many remarkable progress and novelty for industrial and academically. Two of the Nobel Prizes have been granted an award on innovation of catalysis due to noteworthy effect on this field. During the researches on this field, the fundamentals and new concepts

B. Coşkuner Filiz (✉)
Science and Technology Application and Research Center, Yildiz Technical University, Davutpasa, Istanbul, Turkey
e-mail: coskuner@yildiz.edu.tr

A. Kantürk Figen
Chemical Engineering Department, Yildiz Technical University, Davutpasa, Istanbul, Turkey

Inamuddin et al. (eds.), *Sustainable Ammonia Production*,
Green Energy and Technology, https://doi.org/10.1007/978-3-030-35106-9_1

such as promoting, poisoning and structure sensitivity have been developed. Efforts for the development of completely new catalyst formulation with new metals or improving the conventional catalytic system have been still going on [1–3]. Several efforts have been done to develop a industrial preferred catalyst a century ago by multi-promoting of iron catalysts [4].

The USA Department of Energy (DOE) organised a day-long meeting "to break a new ground the scientific struggles associated with improving sustainable ammonia production processes in beginnig of 2016. It was focused on discovery of reusable, recyclable, repeatable and highly active catalysts for sustainable ammonia synthesis in DOE Roundtable Report. The substantial carbon footprint of the current industrial process must be addressed to build foundational principles. The design of the next generation of sustainable N_2 reduction catalysts, that are active under ambient conditions, such as homogeneous, heterogeneous, chemical or biological catalysts are alternative to the high temperature–pressure required processes's catalysts [5]. Currently, there is no viable catalytical system has been known that shows all of the requirements as product selectivity, catalytically active, easily scalable and long-lived for sustainable ammonia synthesis. Various types of earth abundant catalysts for ammonia synthesis have been synthesised up to today. Many classifications could be done according to the ammonia synthesis or type of catalyst precursor metal as shown in Fig. 1.

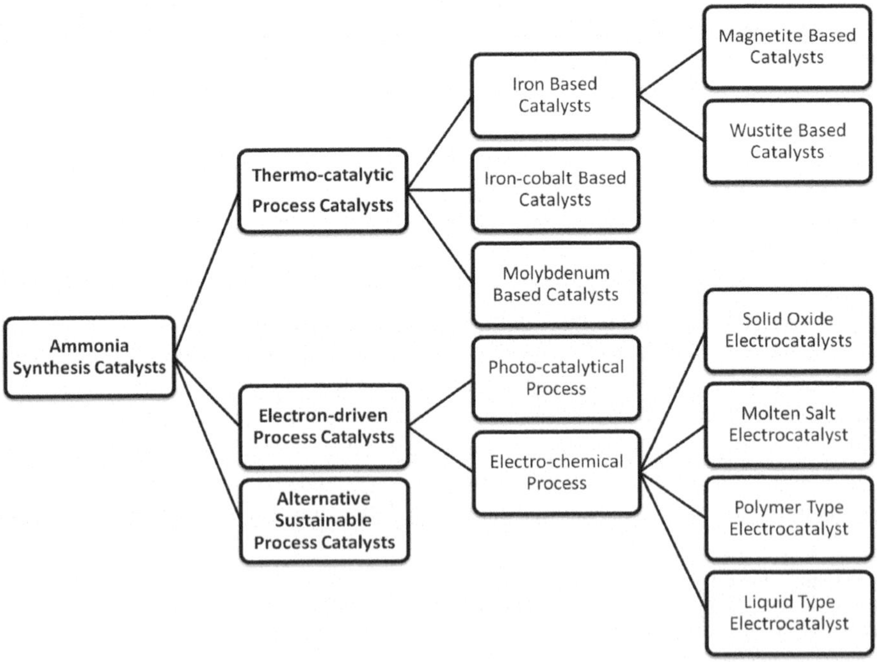

Fig. 1 Classification of ammonia synthesis catalysts according to process

2 Catalysts Based on Process

2.1 Thermocatalytic Process

The thermocatalytic ammonia synthesis has been applied in industrial ammonia production all over the world. The synthesis reaction is based on the interactions between the molecular nitrogen and hydrogen in the presence of catalyst as given below:

$$N_2 + 3H_2 \leftrightarrow 2NH_3 \tag{1}$$

Due to the reaction is reversible and hard to break the highly stable N≡N bond with high binding energy (962 kJ mol^{-1}) in N_2 molecule and also the total volume/pressure of reactants and products decreases with the progress of the reaction, the high temperature and pressure are required for driving the reaction on the ammonia side by increasing the synthesis yield and conversion. Besides the harsh reaction conditions, the most preferred process by industry has been this approach [6–9].

2.1.1 Iron-Based Catalysts

The iron-based catalysts used in thermocatalytic process, as called Haber–Bosch process, that carried out at high pressure (150–350 atm) and temperature (350–550 °C) with low conversion (10–15%) of NH_3 and high energy consumption (485 kJ mol^{-1}) [10–12]. The iron-based catalysts have been investigated for almost a century. These could be classified into two categories as conventional magnetite-based and wustite-based catalysts.

Magnetite-Based Catalysts

The iron-based catalysts firstly prepared from magnetite. Fe_3O_4-based catalysts were deeply investigated for ammonia synthesis, and later on, several researches on improving a new type of catalysts gained importance. The new type of Fe-based catalyst—$Fe_{1-x}O$ (A301 and ZA-5) has been invented (US Patent 5,846,507, 1996; European Patent 0,763,379, 2002; and Germany Patent 69430143T2, 2002.) that gain attraction due to its much lower reduction temperature and higher activity than the conventional magnetite-based catalysts. The catalytic activity of $Fe_{1-x}O$-based ZA-5 and Fe_3O_4-based A110 catalysts were different based on their structural properties: $Fe_{1-x}O$-based catalyst exhibited lower reduction temperature with faster rate. Besides the concave cube-shaped active sites of Fe_3O_4-based catalysts have more (110) plane, the active parts of $Fe_{1-x}O$-based ZA-5 are mixture of sphere and cube with more exposed (211) and (111) planes but less subjected (110) plane [13]. In the iron catalysed systems, the strong adsorption of nitrogen led to negative order, while hydrogen is positive during equilibrium with hydrogen and ammonia [14].

Magnetite-based catalysts are called to be conventional catalysts for ammonia synthesis. They mainly consist of iron oxide Fe_3O_4 and metal oxide promoters as aluminium (Al_2O_3), calcium (CaO) and potassium (K_2O). The type and composition of promoter and Fe^{2+}/Fe^{3+} ratios have been investigated for a long time for ammonia synthesis. According to the classical volcano shape, the appropriate Fe^{2+}/Fe^{3+} ratios were determined to be 0.5 for efficient catalytic activity [15]. Haber–Bosch process industrially used multi-promoted iron catalyst in the past century for ammonia synthesis. Advantages of iron-based catalyst are being high reactivity, a long lifetime and low cost for the ammonia synthesis, yet at high NH_3 concentrations, its overall activity is strongly affected by the decreasing synthesis [1].

Alkali metal is widely used and investigated as promoters in thermocatalytic production of ammonia synthesis. The importance and complexity of promoting have been taken attraction in this field. Aluminium (Al), potassium (K) and calcium (Ca) strongly enrich the surface of catalyst. Oxide forms of Al and Ca are structural promoters from separate particles. Especially, K as an electronic promoter provides good catalytic performance by covering the iron sites and localised active sites/regions on the surface of fused-iron catalyst. Another important factor is distribution of promoters on catalysts surface that affect the reduction and ammonia production [16].

Wustite-Based Catalysts

Wustite-based catalyst discovered by Liu et al. in 1986. This catalyst shows much higher activity and lower temperature of reduction than the conventional magnetite-based catalysts. Several metal promoters such as niobium and cobalt (Co) have been investigated for improving the catalytic properties of Fe-based catalysts. Niobium promotion on wustite-based iron catalyst was enhanced the reducibility of wustite-based catalyst which is highly desired by industry due to decrease time needed for catalyst regeneration and pre-treatment processes. Besides this, niobium promoting inhibited the formation of [2 1 1] plane that is effective for ammonia production [17]. Cobalt addition into wustite catalyst showed a positive catalytic effect on ammonia synthesis reaction. The reduction temperature was shifted to the lower values than wustite catalyst and a reduction rate of catalyst reached the maximum [15].

2.1.2 Iron–Cobalt-Based Catalysts

Cobalt-based catalysts became one of the most investigated ones to develop lower NH_3 inhibition and highly active catalytic systems in recent years, due to cobalt has almost maximum of volcano curve of an intermediate binding energy to nitrogen for NH_3 synthesis similar to Fe and Ru. Despite it is located on the right-handed side that means has lower catalytic activity, higher activation energy and less sensitivity to NH_3 concentration in the gas phase, several strategies such as promoting, alloying and supporting have been applied successfully for improving its activity, general catalytic properties and thermostability [1, 18].

Co-precipitation prepared Co-catalysts exhibited in higher reactivity than the commercial iron catalyst (KM I) [2]. This widely used technique has important parameters as a precursor of an active phase, pH of solutions, a precipitating agent temperature or ageing time that strongly affects the catalysts' properties. The using of different precipitating agents has not been led to a difference in textural properties and active phase surface area for cobalt-based catalysts. Yet using different promoters such as potassium (K), barium (Ba), cerium (Ce) and their combination strongly influences the catalytic activity of the materials for ammonia synthesis [19, 20]. Especially, barium reported being an efficient promoter for Co-based catalysts compared with Fe-based catalysts [1]. Several researchers reported that promoting by a second element such as Ce improved the Co-based catalysts' properties [21, 22].

Raróg-Pilecka et al. reported that promoted unsupported cobalt catalysts with barium and cerium showed more reactivity than the commercial iron catalyst (KM I, H. Topsoe) in ammonia process commonly. The cerium oxide acts as a structural promoter role, inhibiting the sintering of cobalt species and stabilizes the hcp phase of metallic cobalt during process. Also, synergistic relation between barium and cerium improved the catalytic performance of cobalt catalyst. Barium promoting is not only improved the activity but also modified the active phase in the cobalt oxide catalysts [2, 21, 22]. Potassium promoting reported to being decreasing effect of Co/CeO_2 catalyst by altering the adsorption performance of hydrogen and nitrogen molecules, rather than its electronic property [19].

Besides this, using different promoting procedures affects the catalyst's activity. Ce promoting by co-impregnation technique showed better dispersion on catalytic material—Co–Ba/C catalysts that have higher activity compared with subsequent impregnation prepared one [18].

The advantage of the promoting of Co-based catalysts by Ba and Ce elements is being less inhibition by ammonia than the commercial magnetite catalysts [18]. It must be noted that also the amount of promoter is an important factor for improving the catalytic performance; otherwise, higher promoting quantities on the catalyst structure could be results with lower activity [19].

Cobalt molybdenum nitride (Co_3Mo_3N) was indicated to be potentially the more active catalyst due to the active sites for a number of metals than industrial iron catalysts and promoted ruthenium for ammonia synthesis [4, 23]. The promoting by alkali metals such as Ce, K in optimal concentration optimum improved the catalytic behaviour. Ce promotion provides stability under process conditions [23].

The low dispersion of Co metal in catalytic materials resulted in low rate of reaction with respect to metal mass [22]. The improving of dispersion has been increased by using supporting materials such as carbon [18] and cerium (CeO_2) [19]. Not only the type of the supporting materials affect the activity, but also morphology of the same crystalline structured support does the same. Polyhedral, nano-rod and hexagonal-shaped CeO_2 supports have altered the oxidation catalyst content that resulted to altering in reactivity. The highest ammonia synthesis activity was obtained by using polyhedral CeO_2-support which led to higher concentration of Ce^{+3} and lower binding energy of Co species [24].

2.1.3 Molybdenum-Based Catalyst

The nitride form of molybdenum as a non-noble metal is stable under ammonia synthesis conditions that make it a simple material to improve the highly active catalyst [25]. According to volcano-shaped relation of turnover frequency–nitrogen adsorption energy curve of several catalysts for ammonia synthesis, solely Mo-catalyst showed lower activity than the Fe-, Ru- and Co-based catalyst and higher activity than the Ni-based ones. Using combination of Mo and other metal in catalyst structure was suggested to create active sites for desired activity. Mo metal is too strongly binding with N, Co metal is vice versa. This combination was developed to obtain better activity than Fe and Ru catalysts [26]. The bimetallic forms of these catalysts with iron, cobalt or nickel metals performed improved performance as Co–Mo > Fe–Mo > Ni–Mo. Besides these, their ammonia synthesis performance decreased significantly during a prolonged run [27]. Molybdenum nitride catalysts such as Co_3Mo_3N, Ni–Mo–N, Fe–Mo–N and Mo_2N had almost same activation energy values. The addition of a Co, Ni or Fe component into catalyst changed the catalytic activity. The increase in pressure resulted in increased in ammonia production rate similar to iron catalysts, while Cs-promoted Co_3Mo_3N catalyst showed higher activity than Fe–K_2O–Al_2O_3 especially under high pressure [28]. Alkali addition on molybdenum catalyst led to decrease the surface area of Co_3Mo_3N, yet led to development active catalyst for ammonia synthesis [25]. Small amount of Cs promoting in Co_3Mo_3N catalysts proved to have higher activity than that of the commercial multi-promoted iron catalyst [4].

The new catalysts proposed and worked at non-Haber–Bosch conditions. It was mentioned that it was theoretically possible to synthesise the most active catalyst at low pressure and temperature Haber–Bosch process. They discussed the new classes of catalyst materials as transition metal compounds by calculating transition state energies. They calculated the dissociative chemisorption energies for N_2, CO, O_2 and NO on the (110) stoichiometric surface of oxides, and it is found that this class of materials is up and coming for ammonia synthesis if the oxide with ideal binding energy of nitrogen could be discovered [29].

2.2 Electron-Driven Process

Ammonia synthesis by the electron-driven process is based on the reduction of nitrogen molecules by using hydrogen or water molecules to ammonia as given below in Eqs. (2) and (3):

$$N_{2(g)} + 3H_{2(g)} \rightarrow 2NH_{3(g)} \quad (2)$$

$$N_{2(g)} + 3H_2O_{(l)} \rightarrow 2NH_{3(g)} + O_{2(g)} \quad (3)$$

Due to H_2 production is also over costing process, water is more promising hydrogen source. The reactions are strong endothermic and six electrons along with six protons are needed to transfer to formation of ammonia molecules. The catalysts in the field of electron-driven processes mainly classified into two sections: Photocatalytic and electrochemical ammonia synthesis.

2.2.1 Photocatalytic Process

Photocatalytic process for ammonia synthesis is based on the redox reactions. The semiconductors and metal clusters are components of the system that use as photocatalysts in the process. The metal active sites provide dissociative and associative adsorption of reactants and play curial role in improving the photocatalytic activity. Au, Ag, Cu, etc. metals with large free electron density could be easily excited by visible light and brought out collective vibration to form holes and electrons in metal structure. If energy of electron passes beyond Schottky barrier height between semiconductor and metal, electrons transfer to the semiconductor and ammonia synthesis is successfully completed.

The general classification of this class of catalysts could be done as several metal oxides, oxyhalides, graphitic–carbon, transition metal sulphides based catalysts [30]. TiO_2, Fe_2O_3, Fe(O)OH, CuO, WO_x, Sm_2O_3, Ti-exchanged zeolite, V_2O_3, GaP, Bi_5O_7I, etc., several semiconductors have been improved as photocatalysts. The first investigation on the semiconductor types was started with TiO_2 and the phase of rutile was found to be more active [31]. The high temperature heat treatment and promoting by using noble metal oxides such as iron made the photocatalytic behaviour of TiO_2 more active due to creating the impurity states or defects in the band gap of material [32, 33].

The facet design, vacancies design, doping, structure design and co-catalysts design strategies have been developed to obtain more efficient synthesis [31].

The facet design strategy includes bismuth oxyhalides, two-dimensional (2D) materials, ultrathin (UT) materials and layered double hydroxide (LDH) with ultrathin layered structure developments. The UT design provides facile synthesis performance by creating defect-abundant structures in materials that affect the electronic structure and lead to metal–metal electron transfer [34].

The vacancies design approach grounds on creating vacancies in materials structure to generate electrons and donating them to the N_2 molecules. Bismuth oxyhalides, TiO_2 and ternary metal sulphides have been investigated up to today. Bismuth oxyhalides (BiOX, X: Cl, Br or I) as a 2D semiconductors have been taken an attention of having good photocatalytic activity due to easily generation of vacancies in their structure. The only obstacle prevents the sustainability of the material is easy oxidisation of the bismuth oxyhalides during the process [35].

The doping of photocatalysts by metal additives has been useful for promoting the ammonia synthesis. Cu, Mo, Cr, Fe, Ce, V, Mg, Al, etc. elements have been used for adsorbing and activating the nitrogen molecules to transferring the electrons to the adsorbates [36–42].

One of the other strategies to design effective photocatalysts is structure design. Transition metal has been used in co-catalysts design to enhance catalytic activity by promoting the fine-tuning adsorption and electron–hole separation such as homogeneously disperse metal structures supported on the surface of photocatalysts as graphene and TiO_2 [43, 44]. Some of the researchers promoted the co-catalysts by addition of oxides such as Al_2O_3 in the catalyst's structure to improve the catalytic features [45].

Transition metal sulphides are one of the other interesting alternatives for ammonia synthesis due to their relatively narrow band gaps in the photocatalysts. Despite the first studies showed very low production yield, the studies are still popular. The first investigated one was CdS and then, by addition of different transition metals such as Zn, Mo, Ni, etc. into structure, the ammonia synthesis yield was aimed to be increased [30]. By modifying the surface vacancies of photocatalytic sites, the process will be improved for sustainable ammonia production.

2.2.2 Electrochemical Process

Electrochemical process to ammonia synthesis has driven an attention due to reducing the energy consumption (20%) and providing a more sustainable and alternative approach compared with Haber–Bosch process. Series of functional materials as solid oxide electrolytes, molten salt electrolytes, polymer-type electrolytes, liquid-type electrolytes have been developing for catalytic electrochemical process [46].

The theoretical researches concern the ammonia production electrochemically at the ambient temperature and pressure. The density functional theory calculations for electrochemically ammonia formation on pristine earth abundant metal electrodes have been investigated. Mo, Fe, Rh and Ru are found on top of the volcano diagrams with the most active surfaces; besides this, hydrogen gas formation will be a competing reducing reaction to the faradaic efficiency. It was shown that early transition metals such as Sc, Y, Ti and Z could yield the formation of ammonia [47]. For more detail, the classification of electrochemical catalytic materials was given and explained below.

Solid Oxide Electrocatalysts

Solid oxide type electrolytes perform ammonia synthesis at high temperatures beyond 500 °C. The solid oxide electrolytrocatalysts could be classified as a perovskite-type (ABO_3), pyrochlore-type ($A_2B_2O_7$) and fluorite-type oxides (AO_2) [48].

The perovskite-type solid oxides are consisted of a rare earth abundant element (A site) such as strontium, barium, lanthanum and a transition metal (B site) such as zirconium, cerium and gallate. Barium cerate oxide ($BaCeO_3$), strontium zirconate ($SrZrO_3$) and lanthanum gallate ($LaGaO_3$) are typical examples for this class of electrolytes. The divalent (Ca^{2+}) and trivalent (Yb^{3+}, Nd^{3+}, Y^{3+}, Gd^{3+}) cations (M site) could be included in perovskite structures, formulated as $AB_{1-x}M_xO_{3-\delta}$. The ammonia synthesis in presence of this oxide system generally occurs at high temperature range as 500–1000 °C. Doping agent as Gd used in the barium cerate structures for increasing the electrical conductivity of the system by several researchers and under atmospheric pressure −480 °C temperature ammonia synthesised.

The pyrochlore-type solid oxides maintain A^{3+} and B^{4+} cations that have a higher value than 1.22 for their radius ratio. The value is lower than the fluorite structure that is obtained during preparation. Pyrochlore structure transform into fluorite at higher preparation temperature as 1370–2230 °C. Despite, the perovskite-type oxides have high proton conductivity in H_2 atmosphere; the pyrochlore types only have this ability at high temperatures. The fluorite-type solid oxides generally have large radius tetravalent cations (A^{4+}) in their composition such as (Ce^{4+}, Th^{4+}, U^{4+}, etc.) [48].

In solid oxide materials, the biggest problem is low catalytic activity due to their proton activity is high at only at higher working temperature. For this reason, the structures with high Faradaic efficiency and high ammonia synthesis rate required to develop a more accurate temperature range (200–500 °C).

Molten Salt Electrocatalysts

Molten salt type electrolytrocatalysts have been grown interested due to mild working temperatures (200–500 °C), wide electrochemical window, thermodynamic stability, low vapour pressure and high ionic conductivity in electrochemical ammonia synthesis. Generally, molten salts form from NaOH, KOH, LiCl, KCl, CsCl, Li_3N and their eutectic mixtures [46, 49, 50]. Depending on the electrodes such as Fe_2O_3, $CoFe_2O_4$, Ni, etc. and their specific properties, ammonia synthesis could be achieved with high rate and lower temperature. In the presence of nano-catalysts in the molten salt, the electrical conductivity and synthesis efficiency were improved to the higher values [51].

Polymer-Type Electrocatalyst

Polymer-type solid systems are generally maintained membranes such as Nafion, sulfonated polysulfone, etc. and catalytic materials. The biggest advantage of this type of materials is lower working temperature (−80 °C) compared to the previous systems. The proton conductivity of membranes was improved by additives. For this aim, Ni, Li … etc. additives used and they affected the whole process in a different way. The ammonia synthesis efficiency needs to improve in these systems, by decreasing the working temperature below 80 °C and prohibit the back decomposition of ammonia to nitrogen which leads to low ammonia synthesis rate [46, 52, 53]. For electrochemical synthesis of ammonia, samaria-doped ceria $Ce_{0.8}Sm_{0.2}O_{2-\delta}$ (SDC) and $SmFe_{0.7}Cu_{0.3-x}Ni_xO_3$ ($x = 0$–3) were sol–gel technique synthesised to be used as cathode. The ammonia synthesis was performed at low temperature–atmospheric pressure by using a Nafion-membrane electrolyte, Ni-doped SDC anode and Ag-Pt paste current collector. The ammonia production rate in cells with several potentials at 25 °C for $SmFe_{0.7}$-x$Cu_{0.3-x}Ni_xO_3$ (SFCN) materials with several Cu:Ni contents as the cathode and at variety of temperatures and 2 V potential were measured. According to the optimisation results, the ammonia evolution rate was obtained in higher values in the case of using $SmFe_{0.7}Cu_{0.1}Ni_{0.2}O_3$ cathode. The highest evolution rate was reached 1.13×10^{-8} mol cm^{-2} s^{-1} under conditions as 80 °C and 2 V [54].

Liquid-Type Electrocatalysts

Liquid-type electrolytes provide ammonia synthesis at atmospheric pressure–room temperature in a more relevant way. Efforts on improving electrocatalysts such as gold, iron, etc. in designed structures have been taken an attraction of several researchers in this field. Several earth abundant metals such as Mo, Co, Fe, Cu, Ni, Ag and Ti, etc. have been studied for this process [55]. By changing of the particle size of electrocatalysts, electrolyte medium, temperature, supporting material or type of active metal; the yield of the process alters. The liquid medium could be acidic or basic depending on the selection of electrocatalysts in these systems. When the air is preferred as a N_2 source in the strong alkali medium such as KOH, NaOH, etc., alkaline material reacted with CO_2 in air and forms carbonates. As a result of this interaction, low ammonia synthesis yield is obtained. This problem could be prohibited by using less concentrated alkali formulations. Also, it was reported that strongly acidic mediums provide highest yields [56]. The performance of liquid-type electrocatalysts affects from temperature change, even they work at low temperature 5–65 °C and have some limits in mass transfer due to this [57]. Similar to Haber–Bosch process, Mo metal shows good catalytic ability for the electrocatalytic ammonia production [58]. Also, transition metal nitrides, such as W, V, Nb and Cr, reported as being most active ones [59, 60]. VN nano-wire/carbon cloth (VN/CC) was developed as a highly active catalyst for reduction reaction of nitrogen under ambient conditions in acidic media for the purpose of making the Haber–Bosch process much sustainable route. They also provide the new strategy to synthesised transition metal nitrides for reaction of nitrogen reduction. The VN/CC nano-wire array was fabricated based on V_2O_5 nitridation and used to form ammonia by the Mars–van Krevelen mechanism. Ammonia yield and faradaic efficiency of system was reported as 3.58% and 2.48×10^{-10} mol s^{-1} cm^{-2}, respectively. In addition to this, catalyst had shown the excellent selectivity and N_2H_4 was not formed [61].

2.3 *Alternative Sustainable Catalysts for Ammonia Synthesis*

Such alternative sustainable catalysts have been trying to improve with high efficiency and recyclability. Up to the day, several efforts have been done not only for industrial-scale but also for laboratory-scale ammonia production. Catalytic transformation of non-noble metal–dinitrogen complexes under ambient conditions is an alternative method for developing more sustainable ammonia synthesis processes as explained in following Eq. (4):

$$\text{Transition metal–dinitrogen complexes} \longrightarrow H_2SO_42NH_3 + \text{non-noble metal species} \quad (4)$$

Protonation of Mo– and W–dinitrogen complexes with inorganic acids leading to ammonia and it is called a Chatt cycle. High valent metal (VI) species and hydrazine ($H_2N–NH_2$) are also obtained as a rather by-product in some cases after the catalytic reaction [62]. Although V, Fe and bimetallic iron–dinitrogen complexes were also used for catalytic transformation, the ammonia production yield is lower compared with Mo and W–dinitrogen complexes [63–65].

Schrock cycle is also established in terms of sustainability for ammonia production on the principles of reduction of dinitrogen using a Mo–dinitrogen complex in relation to a tetradentate triamidoamine ligand as a catalytic material [66]. In addition to this, it was reported several [HIPTN_3N] Mo systems for ammonia production in the presence of electrons and protons at room temperature–pressure. It also mentioned that dinitrogen is reduced at V or Fe sites. Unfortunately, it was difficult to design nonenzymatic V or Fe, because they are less efficient than Mo [67]. The main difference between Chatt and Schrock cycles is the oxidation states of the central molybdenum atom. Oxidation states of Mo(0) to Mo(IV) were proceed in Chatt cycle, when Schrock cycle proceeds between Mo(III) to Mo(VI) oxidation states [68].

3 Conclusion

The catalysts are the most important issue in the ammonia synthesis. Up to today, several efforts have been done not only for improving the process but also for creating new catalytic materials. It cloud be said that the fundamentals and several aspects of the catalysis have been strongly related to ammonia synthesis. Such key terms and knowledge have been introducing in this field thanks to the ammonia production.

The sustainable development of the ammonia synthesis is based on the green practical approaches. The one of the most important ones is catalysis as an operating tool in process. The earth abundant catalysts provide lower capital investment and higher activity for this process. Thermocatalytic process to sustainable approaches, earth abundant metals have been most interesting ones with their catalytic activities. Iron-based catalysts are still popular in the industrial-scale ammonia synthesis today, have been most important catalytic process and earning the Haber Nobel Prize in

1918. Using this type of metals for the gigantic-scale synthesis played a major role in the most chemical process worldwide. By the modification of the structure of wustite-based catalysts, the sustainable production could be provided. Moreover, several efforts need to supply the ammonia synthesis with more sustainable and green approach for meeting the biggest demand of growing society such as fertilizers for improving the crop yields. And, the key point is improving the earth abundant catalysts not only produce ammonia with high yield but also in ambient conditions with good stability.

References

1. Hagen S, Barfod R, Fehrmann R, Jacobsen Claus JH, Teunissen HT, Chorkendorff I (2003) Ammonia synthesis with barium-promoted iron-cobalt alloys supported on carbon. J Catal 214(2):327–335. https://doi.org/10.1016/S0021-9517(02)00182-3
2. Karolewska M, Truszkiewicz E, Mierzwa B, Keopiński L, Raróg-Pilecka W (2012) Ammonia synthesis over cobalt catalysts doped with cerium and barium. effect of the ceria loading. Appl Catal A 445–446:280–286. https://doi.org/10.1016/j.apcata.2012.08.028
3. Tamaru K (1991) In: Jennings JR (ed) Catalytic ammonia synthesis: fundamentals and practice. Plenium, New York
4. Jacobsen CJH (2000) Novel class of ammonia synthesis catalysts. Chem Commun 12:1057–1058. https://doi.org/10.1039/b002930k
5. Nørskov J, Chen J, Miranda R, Fitzsimmons T, Stack R (2016) Sustainable ammonia synthesis—exploring the scientific challenges associated with discovering alternative, sustainable processes for ammonia production. DOE Roundtable Report. Dulees, VA. https://doi.org/10.2172/1283146
6. Van Der Ham CJM, Koper MTM, Hetterscheid DGH (2014) Challenges in reduction of dinitrogen by proton and electron transfer. Chem Soc Rev 43(15):5183–5191. https://doi.org/10.1039/c4cs00085d
7. Kitano M, Inoue Y, Yamazaki Y, Hayashi F, Kanbara S, Matsuishi S, Yokoyama T, Kim SW, Hara M, Hosono H (2012) Ammonia synthesis using a stable electride as an electron donor and reversible hydrogen store. Nat Chem 4(11):934–940. https://doi.org/10.1038/nchem.1476
8. Li XF, Li QK, Cheng J, Liu L, Yan Q, Yingchao W, Zhang XH, Wang ZY, Qiu Q, Luo Y (2016) Conversion of dinitrogen to ammonia by FeN_3-embedded graphene. J Am Chem Soc 138(28):8706–8709. https://doi.org/10.1021/jacs.6b04778
9. Oshikiri T, Ueno K, Misawa H (2016) Selective dinitrogen conversion to ammonia using water and visible light through plasmon-induced charge separation. Angew Chem (International Edition) 55(12):3942–3946. https://doi.org/10.1002/anie.201511189
10. Reynolds JC, Devins SI (2011) FeMoCo central carbon atom. Science 22–1958(November):99. https://doi.org/10.5061/dryad.6m0f6870
11. Service RF (2014) New recipe produces ammonia from air, water, and sunlight. Science 345(6197):610. https://doi.org/10.1126/science.345.6197.610
12. Spatzal T, Aksoyoglu M, Zhang L, Andrade Susana LA, Schleicher E, Weber S, Rees DC, Einsle O (2011) Evidence for interstitial carbon in nitrogenase FeMo cofactor. Science 334(6058):940. https://doi.org/10.1126/science.1214025
13. Zheng YF, Liu HZ, Liu ZJ, Li XN (2009) In situ X-ray diffraction study of reduction processes of Fe_3O_4- and Fe_1-XO-based ammonia-synthesis catalysts. J Solid State Chem 182(9):2385–2391. https://doi.org/10.1016/j.jssc.2009.06.030
14. Ozaki A, Taylor H (2006) Kinetics and mechanism of the ammonia synthesis. Proc R Soc London. Ser A Math Phys Sci 258(1292):47–62. https://doi.org/10.1098/rspa.1960.0174

15. Czekajło Ł, Lendzion-Bieluń Z (2017) Wustite based iron-cobalt catalyst for ammonia synthesis. Catal Today 286:114–117. https://doi.org/10.1016/j.cattod.2016.11.013
16. Engvall K, Holmlid L, Kotarba A, Pettersson JBC, Menon PG, Skaugset P (1996) Potassium promoter in industrial ammonia synthesis catalyst: studies by surface ionization. Appl Catal A 134(2):239–246. https://doi.org/10.1016/0926-860X(95)00206-5
17. Han W, Huang S, Cheng T, Tang H, Li Y, Liu H (2015) Promotion of Nb_2O_5 on the wustite-based iron catalyst for ammonia synthesis. Appl Surf Sci 353:17–23. https://doi.org/10.1016/j.apsusc.2015.06.049
18. Karolewska M, Truszkiewicz E, Wścisel M, Mierzwa B, Kępiński L, Raróg-Pilecka W (2013) Ammonia synthesis over a Ba and Ce-promoted carbon-supported cobalt catalyst. Effect of the cerium addition and preparation procedure. J Catal 303:130–134. https://doi.org/10.1016/j.jcat.2013.03.005
19. Lin B, Liu Y, Heng L, Ni J, Lin J, Jiang L (2018) Effect of barium and potassium promoter on Co/CeO_2 catalysts in ammonia synthesis. J Rare Earths 36(7):703–707. https://doi.org/10.1016/j.jre.2018.01.017
20. Tarka A, Zybert M, Kindler Z, Szmurło J, Mierzwa B, Raróg-Pilecka W (2017) Effect of precipitating agent on the properties of cobalt catalysts promoted with cerium and barium for NH_3 synthesis obtained by co-precipitation. Appl Catal A 532:19–25. https://doi.org/10.1016/j.apcata.2016.11.030
21. Raróg-Pilecka W, Karolewska M, Truszkiewicz E, Iwanek E, Mierzwa B (2011) Cobalt catalyst doped with cerium and barium obtained by Co-precipitation method for ammonia synthesis process. Catal Lett 141(5):678–684. https://doi.org/10.1007/s10562-011-0564-8
22. Raróg-Pilecka W, Miśkiewicz E, Kowalczyk Z (2008) Activated carbon as a template for creating catalyst precursors. unsupported cobalt catalyst for ammonia synthesis. Catal Commun 9(5):870–873. https://doi.org/10.1016/j.catcom.2007.09.014
23. Moszyński D, Jedrzejewski R, Ziebro J, Arabczyk W (2010) Surface and catalytic properties of potassium-modified cobalt molybdenum catalysts for ammonia synthesis. Appl Surf Sci 256(17):5581–5584. https://doi.org/10.1016/j.apsusc.2009.12.150
24. Lin B, Qi Y, Wei K, Lin J (2014) Effect of pretreatment on ceria-supported cobalt catalyst for ammonia synthesis. RSC Adv 4(72):38093–38102. https://doi.org/10.1039/c4ra06175f
25. Kojima R, Aika KI (2001) Cobalt molybdenum bimetallic nitride catalysts for ammonia synthesis: Part 1. Preparation and characterization. Appl Catal A 215(1–2):149–160. https://doi.org/10.1016/S0926-860X(01)00529-4
26. Clausen BS, Bahn S, Dahl S, Logadottir A, Nørskov JK, Jacobsen CJH (2002) Catalyst design by interpolation in the periodic table: bimetallic ammonia synthesis catalysts. J Am Chem Soc 123(34):8404–8405. https://doi.org/10.1021/ja010963d
27. Mittasch A, Frankenburg W (1950) Early studies of multicomponent catalysts. Adv Catal 2(C):81–104. https://doi.org/10.1016/S0360-0564(08)60375-2
28. Kojima R, Aika K (2001a) Cobalt molybdenum bimetallic nitride catalysts for ammonia synthesis. Part 2. Kinetic study. Appl Catal A General 218:121–28. http://ac.els-cdn.com/S0926860X01006263/1-s2.0-S0926860X01006263-main.pdf?_tid=a07c1d30-2eab-11e7-a3b3-00000aacb35d&acdnat=1493670212_a16b7a35460416536dd1217446faa176
29. Vojvodic A, Medford AJ, Studt F, Abild-Pedersen F, Khan TS, Bligaard T, Nørskov JK (2014) Exploring the limits: a low-pressure, low-temperature Haber-Bosch process. Chem Phys Lett 598:108–112. https://doi.org/10.1016/j.cplett.2014.03.003
30. Xue X, Chen R, Yan C, Zhao P, Hu Y, Zhang W, Yang S, Jin Z (2018) Review on photocatalytic and electrocatalytic artificial nitrogen fixation for ammonia synthesis at mild conditions: advances, challenges and perspectives. Nano Res 12(1). https://doi.org/10.1007/s12274-018-2268-5
31. Wang K, Smith D, Zheng Y (2018) Electron-driven heterogeneous catalytic synthesis of ammonia: current states and perspective. Carbon Resour Convers 1(1):2–31. https://doi.org/10.1016/j.crcon.2018.06.004

32. Bourgeois S, Diakite D, Perdereau M (1988) A study of TiO_2 powders as a support for the photochemical synthesis of ammonia. React Solids 6(1):95–104. https://doi.org/10.1016/0168-7336(88)80048-2
33. Schrauzer GN, Guth TD (2016) Cheminform abstract: Photocatalytic reactions. 1. Photolysis of water and photoreduction of nitrogen on titanium dioxide. Chemischer Informationsdienst 9(6):7189–7193. https://doi.org/10.1002/chin.197806026
34. Zhao Y, Zhao Y, Waterhouse Geoffrey IN, Zheng L, Cao X, Teng F, Li Zhu W, Tung CH, O'Hare D, Zhang T (2017) Layered-double-hydroxide nanosheets as efficient visible-light-driven photocatalysts for dinitrogen fixation. Adv Mater 29(42):1–10. https://doi.org/10.1002/adma.201703828
35. Wang S, Hai X, Ding X, Chang K, Xiang Y, Meng X, Yang Z, Chen H, Ye J (2017) Light-switchable oxygen vacancies in ultrafine Bi_5O_7 Br nanotubes for boosting solar-driven nitrogen fixation in pure water. Adv Mater 29(31):1–7. https://doi.org/10.1002/adma.201701774
36. Maryin C, Martin I, Rives V, Palmisano L, Schiavello M (1992) Structural and surface characterization of the polycrystalline system CrxO/-TiO_2 employed for photoreduction of dinitrogen and photodegradation of phenol surface area determination and porosity x-ray diffractometry apparatuses and procedures for the Ln. J Catal 134:434–444
37. Hao Y, Dong X, Zhai S, Ma H, Wang X, Zhang X (2016) Hydrogenated bismuth molybdate nanoframe for efficient sunlight-driven nitrogen fixation from air. Chem Eur J 22(52):18722–18728. https://doi.org/10.1002/chem.201604510
38. Hu S, Xiaoyu Q, Bai J, Li P, Li Q, Wang F, Song L (2017) Effect of Cu(I)-N active sites on the N_2 photofixation ability over flowerlike copper-doped g-C_3N_4 prepared via a novel molten salt-assisted microwave process: the experimental and density functional theory simulation analysis. ACS Sustain Chem Eng 5(8):6863–6872. https://doi.org/10.1021/acssuschemeng.7b01089
39. Ileperuma OA, Tennakone K, Dissanayake WDDP (1990) Photocatalytic behavior of metal doped titanium-dioxide—studies on the photochemical-synthesis of ammonia on Mg/TiO_2 catalyst systems. Appl Catal 62:L1–L5
40. Palmisano L, Augugliaro V, Sclafani A, Schiavello M (1988) Activity of chromiumion-doped titania for the dinitrogen photoreduction to ammonia and for the phenol photodegradation. J Phys Chem 92:6710–6713
41. Soria J, Conesa JC, Augugliaro V, Palmisano L, Schiavello M, Sclafani A (1991) Dinitrogen photoreduction to ammonia over titanium dioxide powders doped with ferric ions. J Phys Chem 95(1):274–282. https://doi.org/10.1021/j100154a052
42. Tian YH, Hu S, Sheng X, Duan Y, Jakowski J, Sumpter BG, Huang J (2018) Nontransition-metal catalytic system for N_2 reduction to NH_3: a density functional theory study of Al-doped graphene. J Phys Chem Lett 9:570–576
43. Lu Y, Yang Y, Zhang T, Ge Z, Chang H, Xiao P, Xie Y et al (2016) Photoprompted hot electrons from bulk cross-linked graphene materials and their efficient catalysis for atmospheric ammonia synthesis. ACS Nano 10(11):10507–10515. https://doi.org/10.1021/acsnano.6b06472
44. Rao NN, Dube S, Manjubala, Natarajan P (1994) Photocatalytic reduction of nitrogen over (Fe, Ru or Os)/TiO_2 catalysts. Appl Catal B 5(1–2):33–42. https://doi.org/10.1016/0926-3373(94)00042-5
45. Yang Y, Zhang T, Ge Z, Yanhong L, Chang H, Xiao P, Zhao R, Ma Y, Chen Y (2017) Highly enhanced stability and efficiency for atmospheric ammonia photocatalysis by hot electrons from a graphene composite catalyst with Al_2O_3. Carbon 124:72–78. https://doi.org/10.1016/j.carbon.2017.07.014
46. Guo X, Zhu Y, Ma T (2017) Lowering reaction temperature: electrochemical ammonia synthesis by coupling various electrolytes and catalysts. J Energy Chem 26(6):1107–1116. https://doi.org/10.1016/j.jechem.2017.09.012
47. Skúlason E, Bligaard T, Gudmundsdóttir S, Studt F, Rossmeisl J, Abild-Pedersen F, Vegge T, Jónsson H, Nørskov JK (2012) A theoretical evaluation of possible transition metal electrocatalysts for N_2 reduction. Phys Chem Chem Phys 14(3):1235–1245. https://doi.org/10.1039/c1cp22271f

48. Glerup M, Nielsen OF, Poulsen FW (2001) The structural transformation from the pyrochlore structure, A2B2O7, to the fluorite structure, AO2, studied by raman spectroscopy and defect chemistry modeling. J Solid State Chem 160(1):25–32. https://doi.org/10.1006/jssc.2000.9142
49. Giddey S, Badwal SPS, Kulkarni A (2013) Review of electrochemical ammonia production technologies and materials. Int J Hydrogen Energy 38(34):14576–14594. https://doi.org/10.1016/j.ijhydene.2013.09.054
50. Murakami T, Nishikiori T, Nohira T, Ito Y (2003) Electrolytic synthesis of ammonia in molten salts under atmospheric pressure. J Am Chem Soc 125(2):334–335. https://doi.org/10.1021/ja028891t
51. Garagounis I, Kyriakou V, Stoukides M, Vasileiou E, Vourros A (2016) Progress in the electrochemical synthesis of ammonia. Catal Today 286:2–13. https://doi.org/10.1016/j.cattod.2016.06.014
52. Amar IA, Lan R, Petit Christophe TG, Tao S (2011) Solid-state electrochemical synthesis of ammonia: a review. J Solid State Electrochem 15(9):1845–1860. https://doi.org/10.1007/s10008-011-1376-x
53. Xu G, Liu R (2009) $Sm_{1.5}Sr_{0.5}MO_4$ (M=Ni Co, Fe) cathode catalysts for ammonia synthesis at atmospheric pressure and low temperature. Chin J Chem 4(20863007):677–680
54. Xu G, Liu R, Wang J (2009) Electrochemical synthesis of ammonia using a cell with a nafion membrane and $SmFe_{0.7}Cu_{0.3-x}$ Ni_xO_3 ($x = 0$-0.3) cathode at atmospheric pressure and lower temperature. Sci China, Ser B: Chem 52(8):1171–1175. https://doi.org/10.1007/s11426-009-0135-7
55. Tsuneto A, Kudo A, Sakata T (1994) Lithium-mediated electrochemical reduction of high pressure N_2 to NH_3. J Electroanal Chem 367(1–2):183–188. https://doi.org/10.1016/0022-0728(93)03025-K
56. Bao D, Zhang Q, Meng FL, Zhong HX, Shi MM, Zhang Y, Yan JM, Jiang Q, Zhang XB (2017) Electrochemical reduction of N_2 under ambient conditions for artificial N_2 fixation and renewable energy storage using N_2/NH_3 cycle. Adv Mater 29(3). https://doi.org/10.1002/adma.201604799
57. Spencer ND, Schoonmaker RC, Somorjai GA (1982) Iron single crystals as ammonia synthesis catalysts: effect of surface structure on catalyst activity. J Catal 74(1):129–135. https://doi.org/10.1016/0021-9517(82)90016-1
58. Zhao J, Chen Z (2017) Single Mo atom supported on defective boron nitride monolayer as an efficient electrocatalyst for nitrogen fixation: a computational study. J Am Chem Soc 139(36):12480–12487. https://doi.org/10.1021/jacs.7b05213
59. Abghoui Y, Garden AL, Howalt JG, Vegge T, Skúlason E (2016) Electroreduction of N_2 to ammonia at ambient conditions on mononitrides of Zr, Nb, Cr, and V: a DFT guide for experiments. ACS Catal 6(2):635–646. https://doi.org/10.1021/acscatal.5b01918
60. Abghoui Y, Skúlason E (2017) Onset potentials for different reaction mechanisms of nitrogen activation to ammonia on transition metal nitride electro-catalysts. Catal Today 286:69–77. https://doi.org/10.1016/j.cattod.2016.11.047
61. Zhang X, Kong RM, Huitong D, Xia L, Fengli Q (2018) Highly efficient electrochemical ammonia synthesis: via nitrogen reduction reactions on a VN nanowire array under ambient conditions. Chem Commun 54(42):5323–5325. https://doi.org/10.1039/c8cc00459e
62. Tanabe Y, Nishibayashi Y (2013) Developing more sustainable processes for ammonia synthesis. Coord Chem Rev 257(17–18):2551–2564. https://doi.org/10.1016/j.ccr.2013.02.010
63. Gilbertson JD, Szymczak NK, Tyler DR (2005) Reduction of N_2 to ammonia and hydrazine utilizing H_2 as the reductant. J Am Chem Soc 127(29):10184–10185. https://doi.org/10.1021/ja053030g
64. Leigh GJ, Jimenez-Tenorio M (1991) Exchange of dinitrogen between iron and molybdenum centers and the reduction of dinitrogen bound to iron: implications for the chemistry of nitrogenases. J Am Chem Soc 113(15):5862–5863. https://doi.org/10.1021/ja00015a050
65. Sellmann D, Hennige A (1997) Direct Proof of Trans-Diazene in solution by trapping and isolation of the trapping products. Angew Chem, Int Ed Engl 36(3):276–278

66. Yandulov DV, Schrock RR (2003) Catalytic reduction of dinitrogen to ammonia at a single molybdenum center. Science 301:76–78
67. Yandulov DV, Schrock RR (2014) Catalytic reduction of dinitrogen to ammonia at a single molybdenum center. Science 301(5629):76–78
68. Tuczek F, Horn KH, Lehnert N (2003) Vibrational spectroscopic properties of molybdenum and tungsten N_2 and N_2H_x complexes with depe coligands: comparison to dppe systems and influence of H-bridges. Coord Chem Rev 245(1–2):107–120. https://doi.org/10.1016/S0010-8545(03)00064-X

Reactor Design, Modelling and Process Intensification for Ammonia Synthesis

Ilenia Rossetti

Abstract Different ammonia synthesis processes have been considered, based on commercial converters and the relative synthesis loops. Starting from thermodynamic and kinetic modelling considerations, the attention is focused on reactors' design and on the configuration of the ammonia synthesis converter, as included in the whole plant. Examples of processes operating at different pressure and with different catalytic materials are reviewed. Various options for process intensification and energy saving are also discussed.

Keywords Reactor design · Process design · High pressure converter · Kinetics of ammonia synthesis

1 Introduction

Ammonia production is likely the most important and widespread industrial chemical process since one century. The possibility to fix and activate atmospheric, inert nitrogen opened the possibility to employ it in a variety of compounds as an inexpensive and abundant source.

Ammonia is currently produced as 146 million tons per year worldwide (in 2016 [1]), of which ca. 48 in China, 12 in Russia, 11 in India and 9 in the USA. Ammonia is principally produced through the Haber–Bosch process [2] on a productivity scale that often reaches 1000 t/d, with super-giant plants operating on a capacity of 2–3000 t/d and projects for 4–5000 t/d plants. This technology is one of the major successes of the heterogeneous catalytic process applied at a large scale [3, 4].

Its uses are logically linked to the fact that it is an activated, more reactive molecule with respect to N_2. So, besides its direct transformation into urea, ammonium salts or direct derivatives, it is oxidised to NO_x for the production of nitrites and nitrates. Therefore, it is an indirect way to obtain more oxidised products than elementary nitrogen, given that the direct N_2 oxidation is less favourable.

I. Rossetti (✉)
Chemical Plants and Industrial Chemistry Group, Dipartimento di Chimica, Università degli Studi di Milano, CNR-ISTM and ISTM Unit Milano-Università, via C. Golgi 19, 20133 Milan, Italy
e-mail: ilenia.rossetti@unimi.it

Inamuddin et al. (eds.), *Sustainable Ammonia Production*,
Green Energy and Technology, https://doi.org/10.1007/978-3-030-35106-9_2

The current uses of ammonia are for ca. 85% the production of fertilisers, mainly urea, ammonium phosphate, ammonium nitrate and calcium ammonium nitrate. Ca. 5% of ammonia is oxidised to nitric acid, 5% feeds the polyamides production and the remaining 5% is for other uses.

Looking at the food production in China, a linear proportionality can be found between food production and chemical fertilisers' production [5]. Similar conclusions apply to India [6] and the developed northern countries, e.g. USA, Canada and Europe [7].

Ammonia synthesis occurs through a reversible exothermal reaction (R1) [8]:

$$N_2(g) + 3H_2(g) \rightleftarrows 2NH_3(g) \quad \Delta H^\circ = -46\,\text{kJ}\,\text{mol}^{-1} \tag{R1}$$

Therefore, thermodynamic constraints would suggest operation at low temperature to favour reactants conversion. However, the stability of the reactants, nitrogen in particular, forces to increase the operating temperature in order to achieve detectable and practically relevant productivity for kinetic reasons. Furthermore, a catalyst is used to offer a kinetically reasonable pathway for the reaction at moderate temperature (350–500 °C). Yet, to push the thermodynamics to sufficient yield at such a temperature, high-pressure processes must be developed. To date, pressure higher than 65 bar was industrially used, preferentially higher than 140 bar.

Different materials have been investigated as catalysts for this reaction. As for the industrial application, multiply promoted iron-based catalysts from magnetite (Fe_3O_4) as precursor dominated the scene up to the end of twentieth century [9–13]. Alternative formulations based on wustite ($Fe_{1-x}O$) as iron precursor [11, 14] or bimetallic Fe–Co formulations were proposed [15]. Furthermore, supported Ru catalysts were applied in the Kellogg Advanced Ammonia Process (KAAP), which constituted one of the hugest advancements in ammonia synthesis deployment at the end of last century and stimulated a strong investigation on these new catalytic materials in order to improve their resistance to deactivation and their economic sustainability [16–18]. KAAP is adopted at least in seven plants, sized to produce ca. 2000 t/d of ammonia [8] at pressure as low as 90 bar. The main issues with Ru/C catalysts are (i) the inhibition by H_2, which suggests the use of H_2/N_2 feeding ratio below the stoichiometric composition and (ii) the poor carbon resistance to methanation, the reaction catalysed by Ru that may corrode the C-based support [19–29]. Ru demonstrated more active than Fe and has similar sensitivity to oxygen-containing poisons, but it is less sensitive to sulphur [17, 18].

The search for an active catalyst is of course a milestone for this application. Improving activity means the possibility to achieve higher conversion per pass under the same conditions (i.e. higher revenues at the same operating costs) or to achieve the same conversion per pass at lower pressure (i.e. same revenues at lower operating costs).

However, the catalyst, the catalytic converter and the whole process layout represent a synergistic loop that should be optimised at once in order to improve process efficiency and/or decrease the operating costs: in modern terminology, in the last half century the attention has been focused on process intensification. Thus, besides the improvement of materials performances, reactors and process design have been optimised, to reduce energy and raw materials consumption.

Since the reaction does not pose any selectivity problem, i.e. no by-products or separation issues are relevant, the optimisation criteria rely substantially on the atom economy (from a *well-to-wheel* balance perspective) and energy efficiency. In order to understand the effect of the different process parameters, some details on kinetics and modelling issues are provided in the following paragraph. Further paragraphs focus on the different reactors proposed and on whole industrial process layouts. At last, strategies for heat recovery and process intensification are considered.

2 Kinetics and Modelling Issues

The equilibrium constant of the reaction can be calculated from equilibrium data. A good model for its estimation as a function of temperature was proposed (E1) [30] as follows:

$$\log_{10} K_{\mathrm{eq}} = -\frac{59.9024}{R} + \frac{37656}{RT} - 2.691122 * \log_{10} T - 5.519265 * 10^{-5} T + 1.848863 * 10^{-7} T^2 \qquad \text{(E1)}$$

The equilibrium ammonia concentration for stoichiometric mixtures is exemplified in Fig. 1, which evidences a decreasing maximum ammonia concentration with increasing temperature, according to the exothermal character of the reaction. Furthermore, an increase of pressure favours reactants conversion.

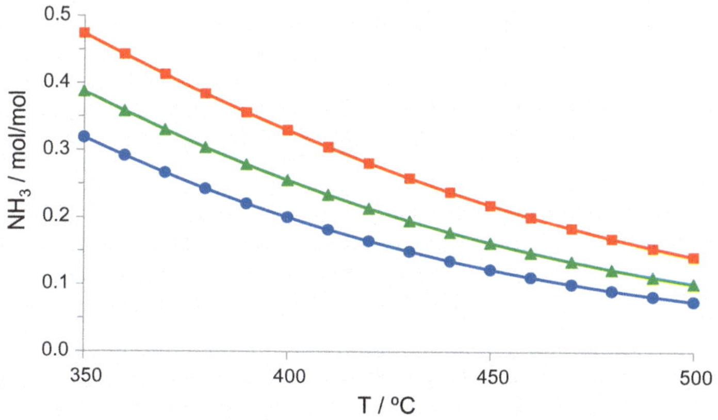

Fig. 1 Equilibrium concentration of ammonia versus temperature with stoichiometric H_2/N_2. Pressure increasing from bottom-up: circles = 70 bar; triangles = 100 bar, squares = 150 bar [31]

From the kinetic point of view, ammonia synthesis is a multi-step heterogeneous catalytic process, in which adsorption/desorption of reactants and product, as well as many intermediate species formed through surface elementary steps, contributes to the complexity of the reaction [30, 32].

$$N_2(g) + 2\sigma \rightleftarrows 2N - \sigma \tag{R2}$$

$$H_2(g) + 2\sigma \rightleftarrows 2H - \sigma \tag{R3}$$

$$N - \sigma + H - \sigma \rightleftarrows NH - \sigma + \sigma \tag{R4}$$

$$NH - \sigma + H - \sigma \rightleftarrows NH_2 - \sigma + \sigma \tag{R5}$$

$$NH_2 - \sigma + H - \sigma \rightleftarrows NH_3 - \sigma + \sigma \tag{R6}$$

$$NH_3 - \sigma \rightleftarrows NH_3(g) + \sigma \tag{R7}$$

The reaction proceeds through a series of adsorption, dissociation, surface reaction and desorption steps [33], involving the surface active sites (σ). The kinetic parameters of each elementary step are a function of the surface coverage and structure. Similar microkinetic mechanism (R2–R7) can be foreseen for either Fe- or Ru-based catalysts, but in most cases R2, i.e. the dissociative N_2 adsorption is considered the rate determining step.

Historically, the best fit of experimental data for Fe-based catalysts was obtained through an original model developed by Temkin [32] (E2), which accounts for the reversibility of the reaction and for the inhibition of the surface active sites by ammonia [10].

A Temkin-type equation has the following formulation:

$$\frac{d\eta}{d\tau} = k\,\lambda(q)\left(\mathrm{Ka}^2 a_{N_2}\left[\frac{\left(a_{H_2}\right)^3}{\left(a_{NH_3}\right)^2}\right]^{\alpha} - \left[\frac{\left(a_{NH_3}\right)^2}{\left(a_{H_2}\right)^3}\right]^{1-\alpha}\right) \tag{E2}$$

where k is the kinetic constant historically referred to ammonia decomposition reaction (rather than the synthesis), Ka is the equilibrium constant, a_i is the activity of species i, and $\lambda(q)$ is adjusted according to the reactants feeding ratio. α is an adaptive parameter, function of the catalyst, to be set at a constant value (normally 0.75) [14, 34].

A genuine Langmuir–Hinshelwood–Hougen–Watson (LHHW) approach did not allow to fully represent catalyst performance for any catalytic system [35]. Therefore, the Temkin equation was modified to extend it to Ru-based catalysts, due to the fact that Fe-based catalysts are kinetically limited by the strong surface adsorption of NH_3, while, on the contrary, Ru is inhibited by competitive adsorption of H_2. Accordingly, a hybrid formulation for a modified Temkin model was developed to better describe Ru catalysts' performance. A sort of mixed Temkin–LHHW approach was used, by adding a denominator to the Temkin rate equation to account for strong hydrogen adsorption, as reported in E3 [34]:

$$\frac{d\eta}{d\tau} = k\,\lambda(q)\frac{\left(a_{N_2}\right)^{0.5}\left[\frac{\left(a_{H_2}\right)^{0.375}}{\left(a_{NH_3}\right)^{0.25}}\right] - \frac{1}{K_a}\left[\frac{\left(a_{NH_3}\right)^{0.75}}{\left(a_{H_2}\right)^{1.125}}\right]}{1 + K_{H_2}\left(a_{H_2}\right)^{0.3} + K_{NH_3}\left(a_{NH_3}\right)^{0.2}} \quad \text{(E3)}$$

where K_a represents the equilibrium constant, $d\eta/d\tau$ the consumption rate of the limiting reactant (depending on feeding stoichiometry) in mol h^{-1} dm_{cat}^{-3}, k is the kinetic constant of the *direct* reaction (not the inverse as in the original Temkin formulation), and a_i the activities of reactants and product. $\lambda(q)$ is again an adaptive parameter depending on the feeding stoichiometry, specifically set for Ru-based catalysts to 1 or 1.2 when the H_2/N_2 feeding ratio is 3 or 1.5.

The equilibrium constants of adsorption to be set at the denominator were calculated from the optimised ΔH_{ads} and ΔS_{ads} for H_2 and NH_3 [34] (E4, E5):

$$\log_e K_{H_2} = -\frac{56.9024}{R} + \frac{37656}{RT} \quad \text{(E4)}$$

$$\log_e K_{NH_3} = -\frac{34.7272}{R} + \frac{29228}{RT} \quad \text{(E5)}$$

Coupling thermodynamics and kinetics requirements for ammonia synthesis reaction allows to predict the effect of the main operating parameters on catalyst performance, so determining the optimal operating points for the plant to be designed. For instance, the isobaric dependence of catalyst performance on contact time τ [here reported as gas hourly space velocity (GHSV (h^{-1}) = $1/\tau$)], needed for reactor sizing, is reported in Fig. 2. Increasing the contact time allows an increase of conversion at the same temperature. These curves present a maximum, which is interpreted as the conditions of maximum rate, which should be selected as the best operating point.

Nevertheless, the reaction is exothermal, so while the reaction is progressing, heat release provokes the increase of catalyst bed temperature, which in turn disfavours the reaction from the thermodynamic point of view. This feature of the reaction suggests different approaches, which will be better described in the next sections.

A first approach would be to remove efficiently the heat released within the catalyst bed. This is hardly accomplished by structuring the reactor itself as a heat exchange reactor and by selecting an appropriate service fluid and heat exchange network. Another option is to quench the reacting system through injection of cold reactants within the bed. A further possibility is to let operate the catalyst bed (or better a portion) adiabatically, subtracting the reaction heat at bed outlet through appropriate heat exchange.

The overall concept is to design a converter operating line, which lies nearby the points of maximum rate and increase step by step the ammonia concentration. This

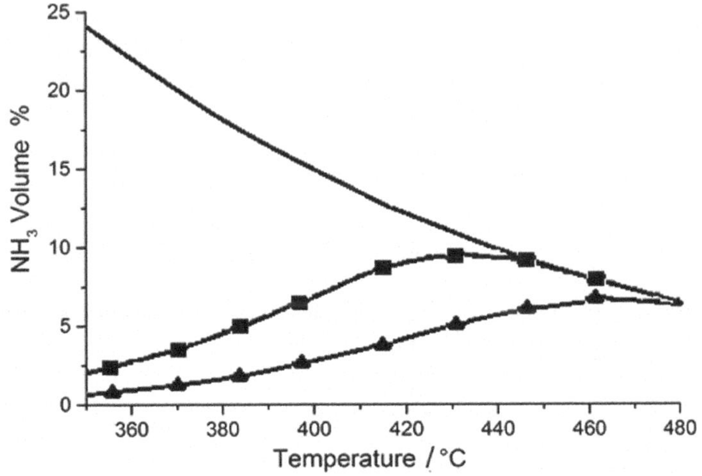

Fig. 2 Equilibrium NH_3 concentration versus temperature (continuous line) compared with concentration predicted from the modified Temkin kinetic model for a Ru/C catalyst at different space velocity. $P = 70$ bar, $H_2/N_2 = 1.5$ v/v, GHSV $= 2.0 \times 10^5$ h^{-1} (triangles) and 0.5×10^5 h^{-1} (squares) [31]

task consists of selecting the most appropriate operating conditions for each catalyst portion (in case constituted by different catalysts types, e.g. in the KAAP option) and achieving suitable heat removal to maintain sufficiently favourable thermodynamic boundaries. Needless to say that, while former ammonia synthesis plants achieved heat exchange by wasting most of the heat subtracted from the converter (only some steam production was coupled), this point is now one of the keys to improve the energy efficiency of the process. Indeed, through this approach, besides preheating the reactants, heat can be made available for other sections of the plant, and high-pressure steam can be produced (mainly in the synthesis gas production section) to operate steam turbines, which in turn drive the compressors to pressurise the synthesis gas. Exhaust steam is further recovered as process heat.

A reliable kinetic model is the primary input for the simulation of reactive systems. Only few reports are available on a full ammonia synthesis process implementing detailed kinetic expressions [31, 34]. The kinetics of NH_3 synthesis was used for the rating of a coal-based process designed to produce synthetic natural gas and ammonia [36]. The control structure of an ammonia synthesis process was studied [37–39] through a Temkin reaction kinetics.

Arora et al. [40] focused on small-scale ammonia synthesis from biomass, but the simulation was based on Gibbs and equilibrium reactors, rather than rate-based models to describe the ammonia reactor. Ammonia production starting from biomass gasification was also investigated [41], but even in this case the converter was represented as a Gibbs-type reactor. The simulation of a rather low-pressure ammonia

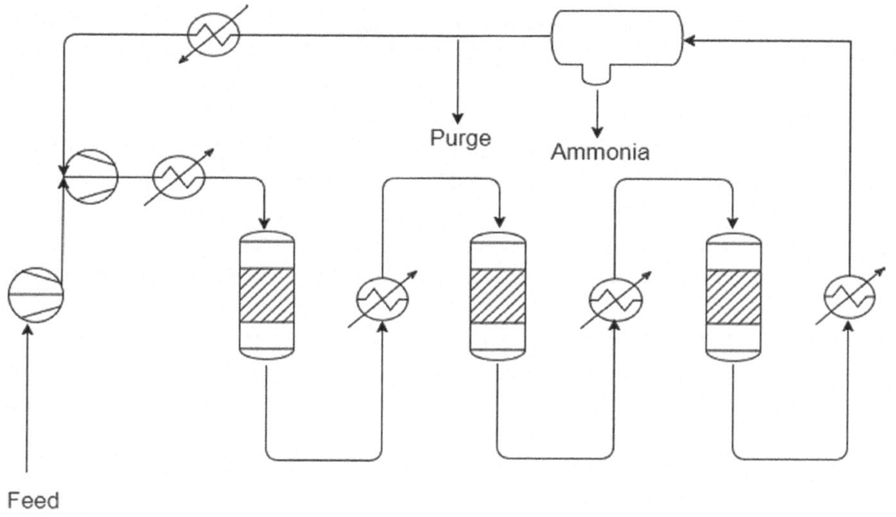

Fig. 3 Sketch of a multi-bed reaction loop used for the simulation of an ammonia synthesis plant [42, 43]. Readapted from [31]

synthesis reactor, operating at 120 bar, was proposed [31], relying on detailed kinetics [14, 34] for a multi-bed converter including different catalyst loading options: Fe from magnetite or wustite or Fe + Ru/C catalyst, as in a KAAP logic (Fig. 3).

Process modelling is achieved by coupling two differential equations, one representing the mass balance in the flow reactor through the kinetic equation and the second computing the temperature profile through a thermal balance. The latter is strictly dependent on the converter design, in particular the heat removal system. Additional equations account for the thermal exchange in heat exchangers [44, 45]. Multi-bed quench converters are also modelled as in the following references [46, 47]. A one-dimensional (1D) isothermal reaction and diffusion model was also developed for the magnetite (A301) catalyst, which is typically characterised by irregular shape (granules). The kinetic equation was obtained fitting data collected from 75 to 105 bar [48].

Optimisation problems as a function of operation parameters are often found. An alternative approach to solve this boundary-value problem and to assess the optimal operation point is shown in [49]. The reactor model coupled with the kinetic equation defines a nonlinear differential-algebraic problem. These differential-algebraic equations are discretized through a finite elements method, so that the resulting algebraic equations are written as equality constraints.

Single and multiple-shooting methods are used with a MATLAB code to solve the ordinary differential equation integration routine [50]. Optimised temperature defines an objective function value of 5.015×10^6 \$/year for an optimised reactor size of 6.695 m. Multiple-shooting method was better than single-shooting one to account for poor initial guesses of the reactor size.

The multiplicity of steady states that may characterise an ammonia synthesis loop is studied from a microkinetic point of view by Mendéz-Gonzàles et al. [51]. In particular, the presence of adjacent surface sites justifies the existence of multiple steady states.

Multivariate optimisation is addressed by Nguyen et al. [52]. The optimal problem requires the maximisation of a multi-variable objective function, including the feed temperature, reactor length and answering to a number of equality constraints. The solution method involved the cyclic search over the coordinate to go through the multi-variable optimisation. A golden section approach allowed to find the maximum value.

Power-to-ammonia loops for use with renewable energy supply characterised by intermittency have been considered, searching for the operating boundaries for steady-state operation of a three-bed autothermal ammonia synthesis reactor through a pseudo-homogeneous model [53].

Finally, the simulation and optimisation of an autothermal ammonia synthesis reactor have been studied [54].

3 Commercial Processes for Ammonia Synthesis

What is conventionally named "ammonia synthesis process" is indeed a complex mix of unit operations, which includes the ammonia converter and ammonia separator, the conditioning of the synthesis gas, i.e. the compression and preheating stage. However, the reactants, hydrogen and nitrogen, must be synthesised in the appropriate feeding ratio, so in common sense, the ammonia synthesis process includes the operations for the synthesis of the reactants starting from raw materials. Nitrogen is supplied through air, while hydrogen can be derived from different feedstocks, currently mostly of fossil source, though some alternatives from renewable sources are now under investigation.

The synthesis gas fed to the ammonia synthesis loop is conventionally constituted of H_2, N_2, CH_4 and Ar, while oxygen-containing compounds are kept in a ppm concentration level because they are poisons for the catalyst. The H_2/N_2 feeding ratio depends on the optimum performance needs of the catalyst, i.e. 3 mol/mol for Fe-based catalysts and ca. 1 mol/mol for Ru-based ones, but also on the optimised operation of the hydrogen production process.

Hydrogen is commonly obtained through different processes starting from light or heavy hydrocarbons or from solid C-containing material, typically coal. Interest is now focused on the use of renewable solid raw materials, e.g. biomass. With natural gas or light hydrocarbons as feedstock, the preferential process is steam reforming, followed by autothermal reforming or partial oxidation, while for heavy oil or coal, partial oxidation or gasification is the preferred routes. The H_2 produced per mole of C consumed (also determining the environmental footprint in terms of CO_2 emissions and atom economy) is different for these processes. When natural gas is economically available, it usually ensures higher efficiency and lower environmental impact.

All these processes lead to a mixture of H_2 and CO/CO_2, plus residual water. These oxygenates are not allowed to contact the ammonia synthesis catalyst, so a deep purification train is a fundamental part of the synthesis gas production and conditioning step. This is achieved through sequential steps of water gas shift, which converts CO into CO_2 according to R8, methanation (R9) or methanolation (R10).

$$CO + H_2O \rightleftarrows CO_2 + H_2 \quad \text{(R8)}$$

$$CO + 3H_2 \rightleftarrows CH_4 + H_2O \quad \text{(R9)}$$

$$CO + 2H_2 \rightleftarrows CH_3OH \quad \text{(R10)}$$

Alternatives or supplementary treatment in some processes is the gas scrubbing with liquid nitrogen, which removes solid particulate (e.g. for coal gasification processes) and condenses most impurities. Older options were based on a copper liquor wash. CO_2 removal is usually accomplished through scrubbing with amines or other specific solvents, or, more recently, through pressure swing adsorption (PSA). Some details on the main units are reported in the following and compared among different industrial processes layouts in Paragraph 5.

3.1 *Steam Reforming*

The reaction occurs through steam addition to a light hydrocarbon (R11 reports the example of CH_4 as feed) as follows:

$$CH_4 + H_2O \rightarrow CO + 3H_2 \quad \Delta H^\circ = 206\,\text{kJ/mol} \quad \text{(R11)}$$

This reaction is strongly endothermal, and it is always accompanied by the water gas shift reaction (R8), which is instead exothermal, so disfavoured at the high temperature of the steam reforming. Nevertheless, some CO_2 is commonly found in equilibrium in the products' distribution. Steam to carbon ratio between 3 and 4 is typically used, but modern plants try to lower as much as possible this ratio to save energy. Feed pre-treatment by desulfurisation is also usually accomplished to prevent poisoning of all the synthesis gas production, purification and ammonia synthesis.

In order to supply heat for the steam reforming, the reaction is typically carried out in tubes host in furnaces, where part of the hydrocarbon feed is burnt. The heating efficiency strongly impacts on the overall hydrogen yield. An adiabatic secondary reformer is usually added in series, where air is admitted to autothermally or exothermally convert the residual methane fraction. The extent of conversion of the primary reformer (temperature and pressure of operation) and of the secondary one (air inlet and adiabatic temperature raise) is optimised for different commercial plants according to various considerations. One is the stoichiometric rather than over-stoichiometric desired ratio between N_2 and H_2. Indeed air addition in the secondary reformer, besides accomplishing almost full methane conversion, represents the point of N_2 inlet in the synthesis loop. If excess air is admitted in the secondary reformer, some adjustment of the synthesis gas composition may be needed.

Typical operating conditions in the primary reformer are 750–820 °C, raising to 950–1025 °C in the secondary reformer, 25–30 bar, using Ni-based catalysts [13].

It should be stressed that from a thermodynamic point of view the steam reforming reaction would be favoured at low pressure. However, the tendency is to increase as much as possible the reforming pressure in order to decrease the pressure gap between the synthesis gas preparation and the ammonia synthesis loop. Indeed, the compression duty between these two sections represents one of the main cost factors. Therefore, the operating pressure of the reforming and purification units should be optimised carefully based on economic considerations on the *whole* process layout.

A methane fraction of 8–12 vol.% (dry basis) exits the primary reformer and is converted in the secondary one, which is operated adiabatically using Ni-based catalysts, converting methane to a residual 0.3–1.0 vol.% (dry basis).

Higher hydrocarbons (e.g. naphta) are usually converted in a pre-reformer located before the main conversion reactor.

3.2 Partial Oxidation

The alternative partial oxidation or gasification are flexible solutions, adaptable to widely different feedstocks, when natural gas is not available or not an economic option.

$$C_nH_m + n/2\,O_2 \rightarrow nCO + m/2\,H_2 \qquad \text{(r12)}$$

The partial oxidation can be applied preferentially to heavy oil or to solid feedstocks, and the main examples are the Texaco and Shell processes. Besides oxygen, some steam is usually added to prevent carbon deposition, and the reaction is operated around 1500 °C, at higher pressure with respect to steam reforming, i.e. 60 bar for the Shell process and 80 bar for the Texaco one. The yield of H_2 per mole of C is lower, but the possibility to operate at such a high pressure has strong advantages on compression costs savings.

3.3 Gasification

It is suitable to treat solid feedstock, mainly coal or lignite [55–57]. It is gaining higher attention currently due to the interest in exploiting solid biomass as a renewable material for hydrogen production.

Commercial processes are classified into low pressure (Koppers-Totzek, mostly used, and Winkler) or high pressure (Lurgi and Texaco). The Koppers-Totzek [58–61] uses powder coal, added with oxygen and steam, at 1550–1650 °C with very short contact time. The Winkler process is based on a fluidised bed operating at ca. 1000 °C [13]. The Lurgi process operates at 30 bar with a moving bed [13, 62], while the Texaco one starts from a coal/water slurry and operates at 40 bar, ca. 1400 °C [63–65].

The mixture leaving the gasifier is quenched with water, which also serves as scrubber to remove carbon particles, then removed from water and recycled to the gasifier.

3.4 CO Removal

All the processes cited above lead to the formation of a $H_2/CO/CO_2$ mixture. CO should be removed till very low concentration (few ppm), and this is efficiently accomplished at first through a water gas shift process (R8).

The reaction is moderately exothermal, but kinetically limited at low temperature. The industrial practice usually prescribes a two-stage treatment using a first reactor to achieve a gross conversion of ca. 90% CO, called high-temperature water gas shift (HT-WGS). The reactor is operated at 350 °C, with a Fe–Cr oxide catalyst [66, 67]. This leaves 1–3 vol.% residual CO in the gas, which is further removed down to 0.1–0.3 vol.% in a second, low-temperature WGS reactor (LT-WGS), operating at 200–250 °C with a Cu-based catalyst [68]. In this way, the HT-WGS step takes advantage of the higher temperature to improve kinetics even with a cheap, not extremely active catalyst. The LT step further refines CO conversion with a much more active, but more expensive catalyst, allowing to decrease the operating temperature so to improve the CO conversion (but operating on a small CO residual fraction) for thermodynamic reasons.

Alternatively, preferential oxidation of CO to CO_2 can be selected, using selective catalysts (e.g. the Selectoxo process [69]).

3.5 CO_2 Removal

The synthesis gas contains up to 20 vol.% CO_2 (dry basis) which can be separated by physical or chemical absorption or by adsorption, which eliminates the major part of this gas. The final removal of the remaining CO and CO_2 is achieved by methanation (R9).

Different commercial processes are available for chemical absorption of CO_2, such as scrubbing with alcanolamines or hot potassium carbonate [12], among which the former is better for the easier and quantitative recovery of the amine in a regeneration tower, which is however one of the main costs of this unit operation [12, 70].

Different solvents have been proposed to decrease the energy consumption and to minimise other drawbacks of the amine process, such as corrosion [71]. Among the physical absorbers, the first choice was water [12, 13], followed by various, more efficient solvents, such as sulfolane + diisopropanol (Sulfinol process [12]), methanol (Rectisol process [72], particularly suitable in case of high CO_2 content) and various mixtures based on polyethylene glycol (Selexol [73, 74] or Sepasolv MPE processes [75]). The advantage of physical absorption with respect to chemical scrubbing is that the regeneration of the solvent is accomplished by depressurisation or stripping, requiring very limited energy input with respect to the heat needed to regenerate chemical absorbents.

3.6 *Final Purification*

Besides a very old option based on "copper wash", the purity level required by ammonia synthesis loops is met by using two options.

One is the cryogenic purification (Braun process [13]), which uses liquid nitrogen to condense excess N_2 added in the secondary reformer and abates contemporaneously CH_4 and Ar. Before the cryogenic unit (ca. −190 °C), the gas should be dried and last traces of CO_2 be removed to avoid freezing, but this leads to a very dry and pure synthesis gas, so to higher energy efficiency of the synthesis loop. The gas exiting this step contains H_2 + 10 vol.% N_2 (additional pure N_2 is needed from air separation units), with ca. 100 ppm of inerts. This low inert composition is particularly interesting for some purge-free synthesis loops, which constitute an economic advantage avoiding the loss of H_2 through the purge. Some gas elimination through purging is instead mandatory in case of higher inerts concentration to avoid their accumulation and consequent progressive dilution of the synthesis gas partial pressure. This option on the other hand needs an air separation plant that guarantees less than 10 ppm O_2.

Another option is the use of PSA to accomplish the final purification (e.g. ICI-LCA process [76]).

Methanation is a third way, by far the most used, which operates at 250–350 °C over a Ni-based catalyst [12, 13], eliminating CO_x below 10 ppm. An alternative is methanolation, which however implies methanol removal by water, its possible separation and recycle as feed to the reformer, and guarantees much lower purity than methanation.

3.7 *Synthesis Gas Compression*

This is one of the main cost items in the whole plant, since the pressure of the synthesis gas must be improved from 25 to 30 bar to the operating pressure of the

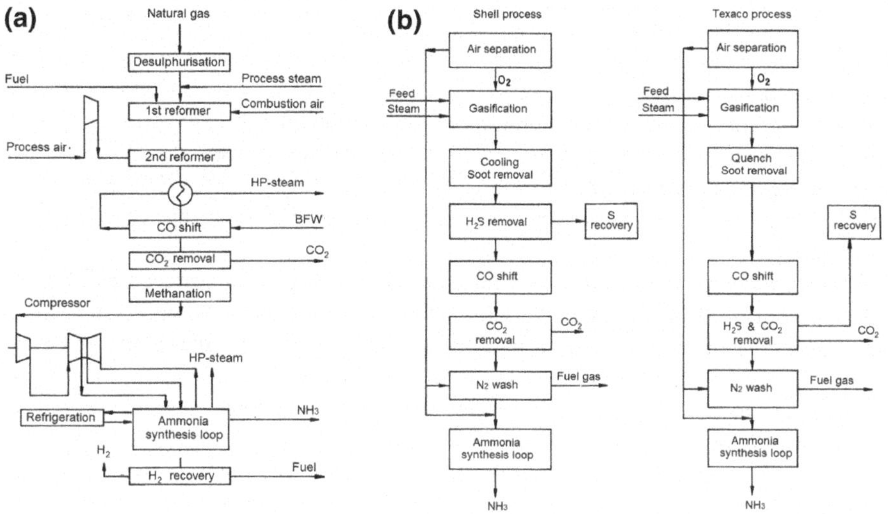

Fig. 4 Examples of block schemes of typical process flowsheets. **a** Steam reforming-based processes; **b** partial oxidation/gasification processes [13]

synthesis loop (widely variable between 150 and 300 bar, with peaks at 80 bar for the Mont Cenis process [77] or 1000 bar in the original Claude process [13]).

The pressure of the synthesis loop, especially in the past, affected the choice of the types of compressors, and, in turn, the latter limited the maximum capacity of the plant. Indeed, very high pressure (e.g. 300 bar) was reachable only by using reciprocating compressors, expensive, with high maintenance cost and poor reliability, needing some further purification of the gas due to contamination with lubricating oil. In the 1960s, centrifugal compressors became available, allowing a considerable increase of capacity from 300–400 to 1000 t/d, however limiting the maximum operating pressure to ca. 150 bar. Improvements from this point of view allowed both an increase of capacity to the current records of 2–3000 t/d and even higher operating pressure to 220 bar. The only disadvantage of centrifugal compressors is their lower efficiency with respect to reciprocating ones.

Examples of block schemes for different integrated processes are reported in Fig. 4 [13].

4 Reactors and Loops Design

4.1 Structure of the Ammonia Synthesis Loop

The ammonia converter is the core or the ammonia synthesis loop, which includes one or in few cases two heterogeneous catalytic reactors, a cooling unit for ammonia recovery with the relative separator, the recycle of the unreacted $H_2 + N_2$, with a make-up and conditioning system. The latter includes preheating and pressurisation to recover the pressure drops across the synthesis loop. The recycle of the unreacted gas is needed due to the low conversion per pass achieved, typically from 12 to 16.5 vol.% NH_3, which can increase to 19–19.5% in advanced processes. An adiabatic temperature rise is expected from 174 to 232 °C (up to 286 °C for the KAAP process), which drives the energy recovery strategy, which can valourise up to 93% of the energy output.

When inerts concentration is relevant, i.e. always except in some cases of cryogenic separation, a purge stream should be withdrawn to prevent their accumulation that causes the progressive dilution of the synthesis gas and intrinsically unsteady operation. The best choice would be to purge at the highest inert concentration, lowest pressure and lowest ammonia content. However, different strategies can take place, strictly depending on the upstream gas processing. Figure 5 exemplifies possible ammonia synthesis loops with their main components. The one reported in Fig. 5A is one of the most used, both in old and modern ammonia synthesis plants (e.g. ICI-AMV, M.W. Kellogg, Braun, Fauser, Linde) [13].

Indeed, the synthesis loop shown in Fig. 5a, with respect to different designs, allows the separation of ammonia before addition of the make-up gas, i.e. at its highest concentration. Furthermore, the purge is withdrawn where the concentration of inerts is the highest and that of ammonia and synthesis gas to be recycled the lowest. The lowest possible gas volume (with minimum ammonia concentration) is sent to the compressors, with consequent energy savings with respect to other options. The purge gas contains up to 30 vol.% of inerts (CH_4 + Ar and some He), some % NH_3, while the remaining is the synthesis gas mixture. Ammonia is recovered from

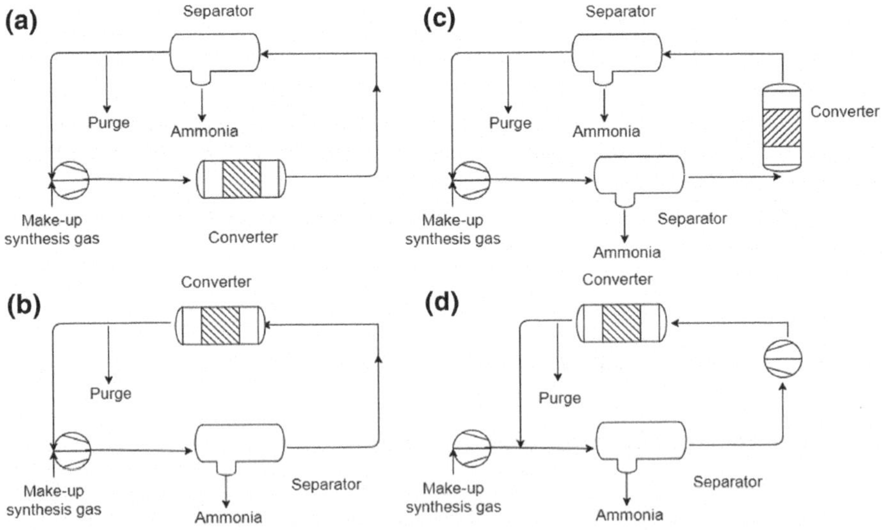

Fig. 5 Examples of ammonia synthesis loops

the purge by scrubbing with water and subsequent distillation, while the remaining gas, containing H_2 is valourised as a fuel, e.g. for the reformer.

Modern plant concepts try to improve hydrogen utilisation by recovering it through membranes (hollow fibres, polymeric membranes), by cryogenic separation or PSA (e.g. Union Carbide, UOP, Linde processes). Metal hydrides such as $LaNi_5$, FeTi or Mg_2Cu can also be used. Hydrogen recovery can be up to 90–95% for membrane technologies, 80–85% for PSA and 90–93% for hydride-based ones.

In particular, ceramic membranes were simulated as permeation devices to achieve hydrogen recovery. A decrease of the gas recycle flowrate up to 8.4% was achieved, reducing the related costs for recompression [78].

Disadvantages of this design reside in the fact that possible purification is made before the addition of fresh make-up gas, with consequent feed to the converter of possible impurities. This latter point is the one that forces in some cases different designs. For instance, an additional purification step of the make-up gas is added before injection into the converter. Sometimes, the passage in the ammonia separator before feeding the make-up gas to the converter is sufficient to ensure abatement of impurities and inerts (Fig. 5b–d).

When the make-up synthesis gas is not very pure, its injection is done before the ammonia separator, referred as three- (Grande Paroisse, Uhde and Topsøe processes) or four-nozzle loops (Casale and M. W. Kellogg processes). This allows the absorption of residual water and CO_2 in the liquid ammonia, but of course this implies a dilution of the mixture to be condensed and separated (lower ammonia partial pressure). Furthermore, the purge is diluted and carries away part of the fresh feed [79, 80]. The main consequence is a lower energy efficiency of the loop, to be carefully evaluated balancing it with the advantage of a less demanding upstream purification.

Multiple ammonia separators may be present, especially for pressure over 250 bar. It should be remarked that the choice of high operating pressure allows higher temperature for ammonia recovery, so that cooling water usually suffices for quantitative condensation of ammonia, while low operating pressure implies the adoption of cryogenic units for valuable ammonia recovery (down to −25 °C). Of course, an increase of pressure allows higher conversion per pass, and hence, lower recycle flowrate, but at the expenses of a higher power consumption for the compression stage. Careful economic optimisation is needed to contemplate all these features. An economic optimum has been found around 155 bar operating pressure.

Moreover, a less efficient ammonia separation implies lower recovery and its higher recirculation in the converter which is unfavourable from a thermodynamic point of view.

4.2 Converter Design

The converter holds the catalytic bed(s), but also accomplishes the most efficient heat exchange, since it was already stressed the need of effective removal of the heat of reaction. The goal is to allow each catalyst to operate near the optimum reaction temperature, i.e. at the highest allowed reaction rate. Meanwhile, the pressure drop across the catalyst bed should be as low as possible. Very different arrangements are possible, operating between 12,000 h^{-1} for plants operating at 150 bar and 35,000 h^{-1} when operating at 800 bar. The temperature ranges between 350 and 530 °C, which is often considered the upper limit for catalyst resistance.

A first option prescribes a direct heat exchange with cooling pipes located within the catalyst bed, or alternatively (but less used) a shell and tube heat exchanger reactor with the catalyst inside the tubes and coolant in the shell. Co-current, counter-current or cross-flow arrangements are possible, and the coolant is the fresh synthesis gas to be preheated or boiler feed water. This arrangement is mainly used in small–medium converters and leads to difficult temperature control with slow dumping ("hunting") [9].

An alternative is the use of cold feed to quench the catalyst, distributing the quencher inside the bed through distribution nozzles or between two adjacent adiabatic beds.

A third option is to run the catalyst beds adiabatically and to remove the reaction heat at once at bed exit, through inter-bed heat exchangers.

Mixed arrangements are also possible.

It is necessary to cope with the very high operating pressure, relatively high temperature, in the presence of hydrogen and nitrogen, which may embrittle most construction materials due to corrosion phenomena, in particular nitriding, hydrogen attack and stress corrosion. Alloy steel can be safely used below 400 °C; whereas stainless steel, Incoloy and Inconel materials should be used for higher temperature. To cope with these points, the so-called *cold-wall* arrangement is usually adopted. The hot side containing the catalyst bed, often called the *cartridge*, is constituted of an internal vessel, around which the inlet fresh feed flows (after moderate preheating, e.g. at 200 °C). The wall of this internal basket is subject to high temperature and is in contact with the synthesis gas, but experiences a small pressure gradient, corresponding essentially to the pressure drop in the loop (5–10 bar). By contrast, the highest pressure gradient is in charge of the external pressure vessel, which is however in contact with a relatively cold fluid, thus allowing less expensive materials.

Different designs are possible, as exemplified in the following. Figure 6 represents a converter where the reactants are split in multiple entrance points and flow in axial direction through the catalyst. A heat exchanger is located in the bottom of the converter, and a complex flow of the synthesis gas accomplishes internal cooling of the catalyst bed. In particular, the coolant flows inside tubes, which is a more common configuration than hosting catalyst in the tubes (Fig. 6, left). An example of this configuration is the TVA converter, used commonly in small size plants, up to 300 t/d (e.g. Topsøe and ICI processes). SBA converters and, more recently, the ICI-LCA ones use a counter-current cooling fluid flow always through cooling pipes and with a cold shell.

A co-current arrangement (Fig. 6, right) is instead used in the NEC and Chemico design, which better ensure a profile of temperature that resembles the highest rate one. More efficient heat removal is accomplished at the beginning of the catalyst bed where the maximum temperature gradient and reaction rate are located.

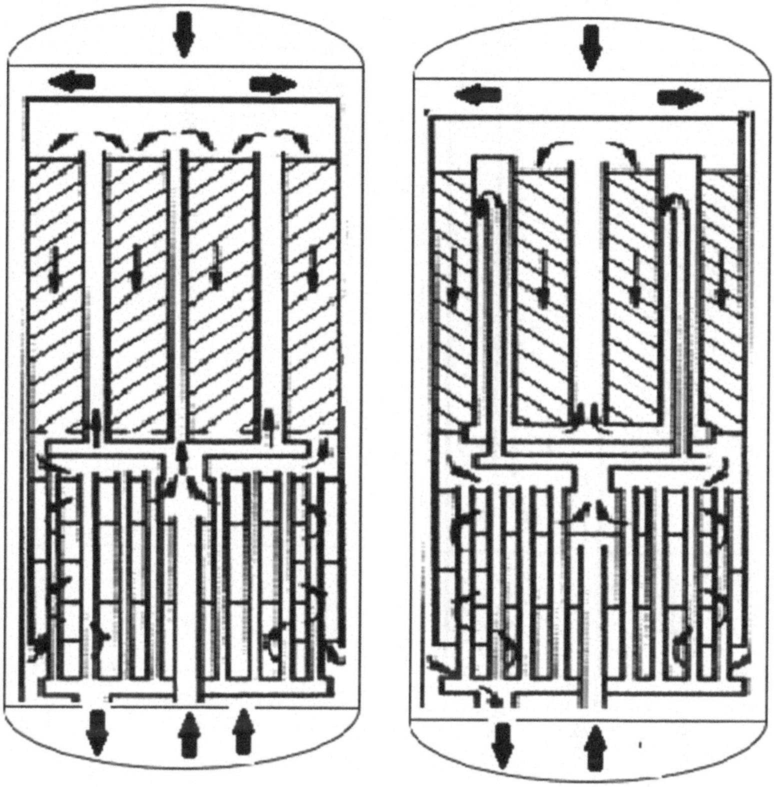

Fig. 6 Cold-wall axial flow converters with internal heat exchanger. Readapted from [13]

The quenching by cold (125–200 °C) feed is exemplified in Fig. 7, left. This arrangement is appropriate for big capacity plants and is very widespread though progressively substituted by indirect cooling which has intrinsically higher efficiency (Fig. 7, right). Examples are the Uhde and the BASF converters. Multiple inter-bed quenching is provided in the Kellogg-type converter (Fig. 8, left), which has a cold-wall design and hosts a conventional heat exchanger on the top. In this way, a zig-zag temperature profile is searched, to lye around the optimal rate operating line. The example proposed on the left of Fig. 8 keeps an axial flow design, while in the right is proposed a conceptually similar heat exchange system, but based on a radial flow.

The step from axial to radial flow configuration was a milestone. Indeed, axial flow converters are typically characterised by higher pressure drop, which can be partially limited by employing bigger catalyst particles. However, the latter are characterised by internal diffusional limitations, and hence, low effectiveness factors. The only alternative is to limit the catalyst bed length by increasing the reactor diameter, which is not straightforward for the design requirements of the high-pressure vessel. Therefore, the conversion into a radial flow design allowed to limit the pressure drop across the catalyst bed without affecting significantly the reactor diameter.

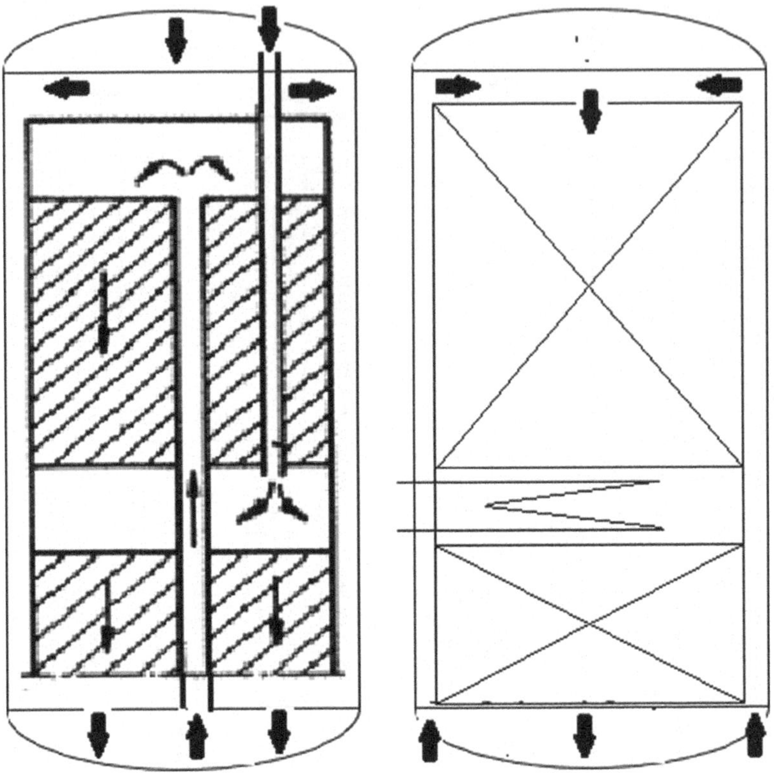

Fig. 7 Cold-wall axial flow converters with quenching (left) or inter-bed indirect heat exchange (right). Readapted from [13]

Much smaller catalyst particles can be used in radial flow converters, even reaching an effectiveness factor equal to 1, i.e. with negligible diffusional limitations. This means full catalyst utilisation, and hence, lower catalyst bed volumes with respect to axial flow converters [81]. Similar designs are proposed by Topsøe in the S-100, S-200 and S-300 converter series (Fig. 9) [82].

A 1D heterogeneous model was applied to an intercooled horizontal reactor, by comparing a solution with two quench flows. Mass, energy and momentum balances have been numerically integrated by Runge–Kutta fourth-order algorithm implemented in a MATLAB code. The simulation results were compared with real plant data with good matching [84, 85].

Hydrodynamic problems encountered in the reconstruction of radial flow configurations have been discussed in [86], where alternative approaches to the design of distributing devices are considered to ensure the radial feed of reagents into the catalyst bed and their uniform distribution.

A mixed axial/radial flow concept has been introduced by Ammonia Casale [87]. This mixed flow pattern is due to multiple gas inlets, in part from the top, which

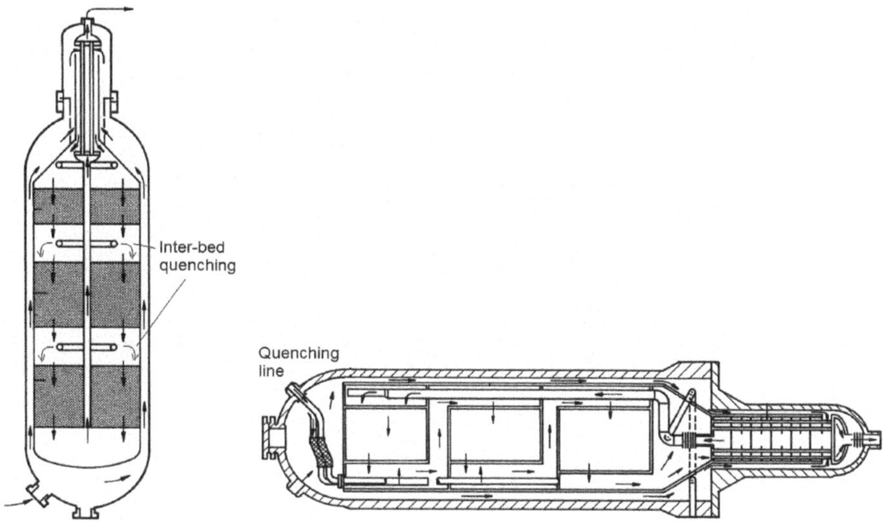

Fig. 8 Design of a Kellogg-type converter with inter-bed quenching. Left: axial flow; right: radial flow. Readapted from [13]

would imply axial downward direction, part from perforation of the co-axial wall, which induces radial direction across the annular catalyst bed. A 2D model has been developed to model an axial–radial ammonia synthesis reactor. Mass and energy balances were solved by a finite volume method, implemented in MATLAB. By contrast, momentum balance and the general continuity equations were handled with COMSOL Multiphysics with a finite element method. The two tools were linked to achieve the simultaneous solution of momentum, mass and energy balance equations. Simulation results were validated with real performance results obtained in a petrochemical plant [88].

A disadvantage of quench cooling is the partial catalyst bed utilisation by the recycle gas, which is injected and allowed to cross only a portion of the catalyst. This increases the catalyst volume with respect to indirect heat exchange, but no extra space is required to allocate the internal heat exchanger. Hence, the total volume of the converter remains almost the same.

Finally, converters with indirect cooling are shown in Fig. 7, right, where heat exchangers placed in the converter, e.g. between catalyst beds, accomplish the removal of the heat of reaction. The cooling fluid can be water, allowing to obtain high-pressure steam, both in axial flow (e.g. Fauser–Montecatini technology [77, 89, 90], ÖSW [13]) or radial flow configuration (e.g. the horizontal Kellogg converter [91, 92], Fig. 10).

The design of the KAAP process, besides introducing an innovative Ru/C catalyst in a multi-bed configuration, also modifies the converter design, going back to a hot-wall design, possible thanks to the much lower operating pressure (ca. 90–100 bar).

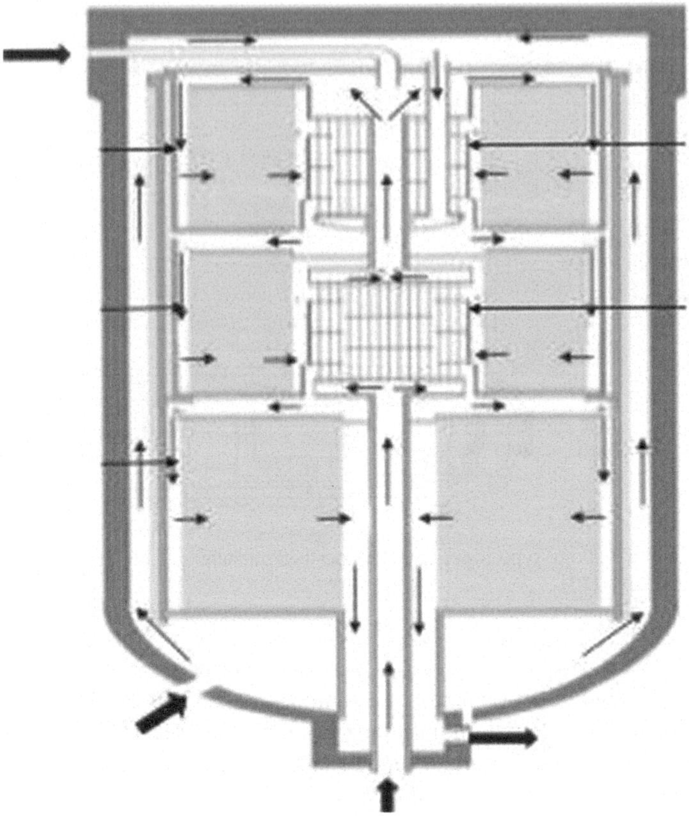

Fig. 9 Design of a Topsøe, three-bed, radial flow, quench-cooled S-300 converter. Readapted from [83]

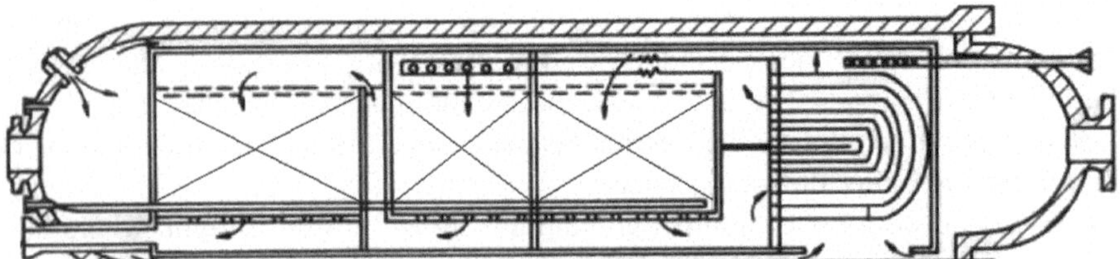

Fig. 10 Horizontal M.W. Kellogg converter with indirect heat exchanger. Readapted from [13]

A horizontal-type industrial ammonia synthesis (Kellogg) with three axial flow catalyst beds and an internal heat exchanger has been also modelled by Dashti et al. [93] through a one-dimensional and non-homogenous model. The optimisation of different ammonia synthesis reactor configurations, with internal direct cooling reactor, adiabatic quench cooling and adiabatic indirect cooling was presented [94].

A 1D pseudo-homogeneous model has been applied to check the role of different parameters on reactor performance and optimisation through a differential evolution algorithm, with N_2 conversion as an objective. Internal direct cooling was more advantageous in the presented case.

Feed-effluent heat exchanger increases thermal efficiency, but could induce hysteresis phenomenon. A double-bed reactor has been dynamically modelled, tuning the quenching ratio to overcome hysteresis [95].

4.3 Ammonia Separation

The separation of pure ammonia is achieved by cooling until liquefaction. For high-pressure synthesis loops, air or water cooling is suitable to lower the residual NH_3 partial pressure in the recycle stream; whereas, for low-pressure cycles, refrigerating systems are needed, down to −25 °C. Liquid ammonia is separated and further flashed to 20 bar to free the absorbed gases, often valourised as additional fuel, and further flashed to near ambient pressure for storage. Possibly, recovery of ammonia from the flash vapour lines is accomplished by scrubbing with water and distillation of the solution obtained.

Alternative options are the recovery of ammonia through scrubbing with dilute ammonia solutions at ambient temperature, leaving a residual ammonia concentration in the loop stream ca. 0.5 vol.% (SnamProgetti) [96]. A further possibility is adsorption on solids [97].

5 Process Intensification

Ammonia synthesis industry accounts for 3% of the world's energy consumption. The revamping of existing plants or the design of new plant configurations from the grass roots answers in most cases the need of higher energy efficiency. This is often accomplished through an intensification of energy recovery to preheat the feed and to produce high-pressure steam, up to 120 bar. Advanced ammonia concepts produce up to 1.5 t of high-pressure steam per ton of ammonia, corresponding to ca. 90% recovery of the reaction enthalpy [9].

The temperature reached in different process steps is reported in Fig. 11, as a basis for the design of an optimised heat exchanger network.

High-level surplus energy is available from the reformer. Steam produced is usually employed in the turbines that drive the synthesis loop compressors.

The development of better catalysts is a valid strategy for process intensification, even in the case of expensive materials. Indeed, the cost of the catalyst is less than 1% of the production cost of ammonia (0.26–0.6% according to [11]), justifying the search for much more active materials. For instance, the KAAP process includes a Ru-based catalyst, which is by far more expensive than the traditional Fe-based one.

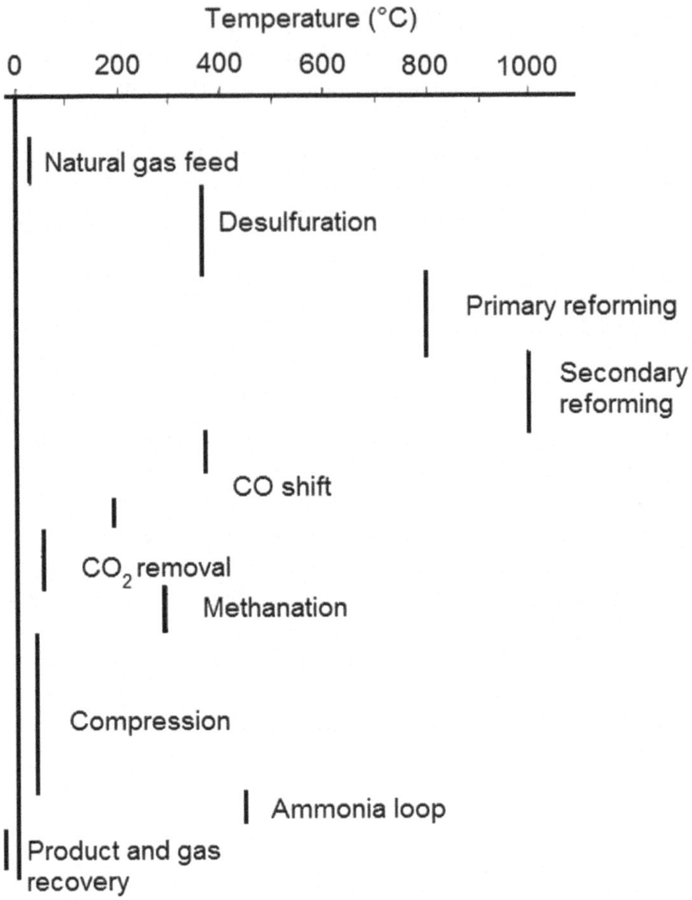

Fig. 11 Temperatures of the different process units. Readapted from [9]

However, the possibility to operate down to 91 bar allows 20% savings in equipment and piping, which justifies fully the higher cost.

Another strategy is often to reduce the H_2/N_2 ratio, to increase the net value of NH_3 in the synthesis loop and save energy in the synthesis gas preparation section.

Moreover, pinch analysis and the optimisation of the heat exchanger network are a must, allowing the recovery and interexchange of most heat flows. Typical examples of steam generation circuit and its use in steam turbines to drive the synthesis loop compressors and of recovery of low-grade heat are reported in Figs. 6.21, 6.22 and 6.3 in [13].

The efficiency of the process can be expressed on the basis of the atom economy, considering the starting raw material. For instance, when hydrogen is produced by steam reforming of natural gas, ca. 0.44 mol of CH_4 is needed per mol of liquid ammonia produced. The "energy efficiency" of the process can be consequently

Table 1 Energy analysis, readapted from [13]

	HHV	LHV	Exergy
Reformer feed	24.64	22.26	23.26
Fuel	7.87	7.07	7.45
Total energy to HP steam production	11.59	11.59	5.52
Losses			
Reforming	0.38	0.38	4.94
Steam generation	0.33	0.33	2.38
Gas purification	1.30	1.30	0.67
Synthesis	1.72	1.72	1.55
Turbines + compressor	6.53	6.53	0.54
Total losses	11.55	10.92	10.54
Energy in NH_3	20.96	17.11	20.13

Values reported in GJ/t NH_3

referred to the lower or higher heating values (LHV or HHV) of the starting material, or by looking at the exergy analysis. The latter approach has the advantage to identify the main plant losses due to irreversibilities, which is a valuable tool to improve the process efficiency. For instance, it was evidenced that most of the exergy loss resides in the reforming section (60–70%), mainly due to the low efficiency of combustion and process steam generation [98]. An example of exergy analysis is reported in Table 1.

A detailed plant (1000 t/d) optimisation based on the minimisation of exergy destruction has been proposed [99]. The proposed objective function resulted very sensitive to some variables, causing convergence issues. Within the synthesis loop, the ammonia converter and the refrigeration were in charge for 71–82% of the total destruction of exergy (38.8 MW for the base case operating conditions, while 25.6 MW for the optimised ones).

Exergy losses were calculated by dividing the process into functional blocks [100]. The reforming block was responsible of the largest exergy loss (3098 kJ/kg). Therefore, this is the block with the hugest potential for intensification. Strategies include preheating the cold feed using an economiser and pressure drop reduction in the secondary reformer. The second widest exergy destruction is represented by the ammonia synthesis loop (3075 kJ/kg). The strategies suggested for optimisation were to limit the temperature increase in the compressor, thermal insulation of the latter, reducing undesired compounds (e.g. Ar) in the feed and decrease the temperature for reactor outlet streams. Exergy losses in the residual ammonia removal unit (833 kJ/kg) were attributed to the stripper and the absorber column, so that improved column design is needed.

The energy consumption for the production of 1 ton NH_3 has been reduced from 41.8 to 29.3 GJ, thanks to substantial energy recovery, integration between different plant sections and optimisation of the operating conditions. In two words, by process intensification. This result has been achieved, e.g. by increasing the steam reforming pressure, to save power during the further gas compression and decreasing the steam to carbon ratio to save process steam, one of the causes of inefficiency. However, this imposes harder conditions for reforming, requiring improved materials and reformer tubes. The matching with a secondary reformer also needs optimisation. The primary reforming can be less demanding, i.e. allowing a decrease of temperature, admitting a higher discharged amount of methane, to be converted in the secondary unit. However, in this way, the secondary reforming operates with higher air flowrate, which requires additional compression power. Additionally, the resulting gas is characterised by higher N_2 amount than required, needing its separation.

The gas purification sections have been improved in the past by adding a low-temperature shift and also CO_2 recovery, by optimising the absorption procedure. For the ammonia synthesis loop, attention has been paid to the design of the converter, to guarantee maximum heat recovery and exploitation. The valourisation of H_2 by recovering it from purge streams also improved the efficiency of the process.

A comparison between the energy consumption of different processes and the relative ratio with respect to the theoretical one has been calculated [11] and summarised in Table 2. The comparison evidences that the most efficient way to accomplish synthesis gas production is the steam reforming of natural gas, especially by increasing its operating pressure, while coal is a low efficiency raw material, even working at high pressure, due to lower hydrogen yield.

Exergy analysis allows to define the links within a system and the intensification potential from a thermodynamic point of view. For instance, two synthesis loop configurations were analysed in [101]. The first scheme includes a three-stage adiabatic reactor, with intermediate quench cooling, while the other design includes a cooled reactor. The exergy destruction could be reduced by enhancing the waste heat utilisation through a steam generator and improving the performance of the

Table 2 Energy consumption for NH_3 production starting from various starting materials (values referred to a ton of liquid NH_3) [11]

Raw material	Synthesis gas production	Capacity (t/d)	Energy consumption (GJ)	Ratio with theoretic energy demand
Natural gas	SR 36 bar	1000	28.4	1.41
	SR 30 bar	1000	35.4	1.76
Light oil	SR 33 bar	900	41.8	2.08
Residual oil	G 85 bar	1000	38.5	1.92
Coal	Pressurised Gasification (Lurgi)	1000	50.4	2.50
	Pressurised Gasification (Texaco)	1000	44.1	2.19

SR steam reforming; *G* gasification

main compressor. Reactor design does not play a key role, since the improvement potential is small. However, it was tackled as the most important component due to its effect on the heat integration.

The use of ammonia as the solar energy storage system has been recently proposed [102]. Concentrated solar energy can be used to thermally drive the endothermal ammonia decomposition and is then recovered during the exothermal synthesis. The heat released is transferred to a fluid that can be superheated steam, or, more efficiently, supercritical CO_2. In order to improve the thermal energy recovery, supercritical CO_2 heating and its use in a Brayton cycle have been investigated in comparison with superheated steam generation to be employed in a conventional Rankine cycle [103]. A bi-annular reactor configuration has been tested, with an outer annulus where synthesis gas preheating is achieved. This is fed to an intermediate annulus hosting the catalyst bed and in the internal pipe, and the heat of reaction is transferred to CO_2. A detailed numerical simulation of ammonia-based thermic cycles is proposed by Abdiwe and Haider [104]. The effectiveness of this configuration is higher than external heat exchange and, besides the very specific application, may suggest next developments to efficiently exploit reaction heat [102, 103, 105].

Next generation catalysts have been developed to improve the conversion per pass, also decreasing the recycle needs, or to decrease the operating pressure at the same productivity. Provided that the activity of a catalyst can be improved at the same volume of about four times, an energy saving of ca. 0.84 GJ/t NH_3 can be achieved. This value increases if a decrease of temperature can be achieved at the same time [11].

The selection of low-pressure operation allows to save energy, but optimisation suggests minimum overall costs for ca. 150 bar. The cost of compression impacts both on installation and operating costs and should be carefully addressed. Typical operating ranges are reported in Table 3. For instance, a lot of low capacity plants are active in China (100–600 t/d) operating at 300 bar, which are currently being revamped to higher capacity and lower pressure (150 bar) to reduce the energy consumption. It should be also reminded that higher operating pressure allows easier ammonia separation, decreasing the refrigeration power.

It is visible that the main power consumption is correlated with the synthesis gas compression, thus there is a tendency to increase the synthesis gas production pressure. This is rather easy in case of solid raw materials, where some commercial gasification processes allow producing the synthesis gas at high pressure (75–95 bar),

Table 3 Examples of centrifugal compressors [11]

Compressor	Natural gas	Air	Synthesis gas and recycle	Ammonia
P-range (bar)	6.8–41	41	27–151	0–18
Power (MW)	1.1	6.3	13.0	7.8

Power expressed for a productivity size of 1000 t/d NH_3

so that the further compression of the gas is no more needed, provided that a low-pressure ammonia synthesis loop is selected. Nevertheless, the great majority of the processes need synthesis gas compression.

The overall energy saving when decreasing the operating pressure was estimated for both small and large capacity plants [11]. For a 1000 t/d plant, the decrease of the loop pressure from 220 to 75 bar allows 1.3 GJ/t energy saving, value which slightly increases for a 300 t/d capacity to 1.38 GJ/t.

Furthermore, the energy consumption of a synthesis loop operating at different pressure with a wustite-based ZA-5 commercial catalyst has been reported for a plant with small capacity (330 t/d, Table 4) [11].

The situation in the 1990s led to only five licensers for proprietary ammonia technologies Braun, ICI, M. W. Kellogg, Haldor Topsøe A/S and Uhde GmbH.

The typical coupling between catalyst choice and process type is summarised in Table 5, while comparison between different processes is provided in Table 6.

At last, due to decreasing catalytic activity during time, an increase of operating pressure is usually planned, also reducing the inert and ammonia inlet to the converter. Alternatively, the increase of the recycle flow is accomplished, which should be taken into account during sizing and, of course, worsens the energy profile of the process. A compromise is searched between the frequency of shut down to replace the catalyst and the admissible loss of activity compensated by worse operating conditions. For small-scale plants, the catalyst life is typically 2–3 years increasing to more than 5

Table 4 Energy consumption calculated for the synthesis loop of a 330 t/d loaded with a ZA-5 catalyst [11]

Pressure (bar)	100	300
Power of the fresh syngas compressor (MW)	7.97	10.4
Power of the circulating compressor (kW)	295	240
Power of the refrigeration compressor (kW)	311	184
Amount of vent gas (m^3/h)	1853	1405
Catalyst volume (m^3)	19.94	5.78
Total power consumption (MW)	13.6	21.6
Specific power consumption (MW/t NH_3)	1.05	1.66

Table 5 Criteria for selection of catalysts and processes [11]

Catalyst	Process	Pressure (bar)	Energy consumption (GJ/t NH_3)
Fe	Small–medium scale	300	29.3
Fe	Kellogg	150	1.4
Fe-Co	ICI-AMV	120	30.1
Ru	Kellogg, KAAP	91	26.8

Table 6 Comparison of different processes for ammonia synthesis [11, 13]

Process	Braun	ICI-AMV Fe–Co catalyst	ICI-LCA	M.W. Kellogg	Kellogg Ru catalyst or Fe + Ru	Uhde	Topsøe	Exxon	Fluor	PARC KTI
H_2 production	Reforming with 50% air excess	Reforming with air excess	Reforming without firing (heat from secondary ref.); large air excess	Reforming with stoichiometric air	Reforming leading to H_2/N_2 = 1 mol/mol	Reforming with stoichiometric air	Reforming with stoichiometric air	Reforming with stoichiometric air	Reforming	Reforming with enriched air
Air compressor	Gas turbine driver; exhaust from turbine used as combustion air in primary reforming	Steam turbine driver; Steam produced in the process	Steam turbine driver			Steam or gas turbine driver	Steam or gas turbine driver	Gas turbine driver	Gas turbine driver	Production of power in a Rankine cycle
CO removal	HT- and LT-WGS	HT- and LT-WGS	First-stage shift (265 °C)	HT- and LT-WGS		HT- and LT-WGS	HT- and LT-WGS	HT- and LT-WGS		HT-WGS
CO_2 removal	MDEA scrubbing (BASF process)	Selexol process	PSA	Selexol		MDEA	Benfield or Vetrocoke; later version: MDEA or Selexol	Catacarb	Polypropylene carbonate	Optional
Final purification	Methanation + dryer + cryogenic (Braun purifier)	Methanation + dryer + cryogenic only on a portion of the gas	Methanation	Methanation + drying with molecular sieves		Methanation	Methanation	Methanation + drying	Methanation	PSA

(continued)

Table 6 (continued)

Process	Braun	ICI-AMV Fe–Co catalyst	ICI-LCA	M.W. Kellogg	Kellogg Ru catalyst or Fe + Ru	Uhde	Topsøe	Exxon	Fluor	PARC KTI
Pressure (kg/cm^2)	150–180	90–120	80	140–180	80–250	160–180	260–140	140		350
Synthesis loop	"Dry", make-up after NH_3 separator. Deep cooling to remove excess N_2, H_2O and CO_2	–		"Dry", make-up after NH_3 separator	With or without synthesis gas recirculation. Improved NH_3 recovery		Recycle compressor after NH_3 sep. Cooling with water + NH_3	"Dry", make-up after NH_3 separator		
Converter	2–3 single bed, adiabatic, axial flow, inter-bed cooling with preheating of the feed and/or steam production	Three bed with quench + indirect cooling	TVA	Two-bed, horizontal, indirect cooling, additional heat exchanger; purge gas recovery	Hot-wall design; multi-bed; radial flow	Two converters	S-200 (radial) or S-250 (radial)	Two-bed radial flow + cryogenic recovery of purge gas		
Net energy consumption (GJ/t NH_3)	28.03	29.29	26.94 (31.80 actual)	29.00	26.78	27.91–30.12	29.16 (27.91 actual)	29.00	37.20	33.93
Capacity (t/d)			450		1850	1500				

for large-scale plants. This means that sizing has to account for widely varying, more and more demanding conditions, especially for large-scale installations. An example is reported in [11], for a plant with a nominal capacity of 1000 t/d, operated at +5% (1050 t/d) for the first 7 years, to account for the loss of performance down to 900 t/d for the next 8 years.

The combination of the ammonia synthesis reaction with some endothermal processes has been investigated at a level of conceptual design. This strategy may be helpful to find the potential to exploit a high revenues product as an economic driver to sustain small-scale, delocalised ammonia production plants.

A new integrated process was designed for a gas-to-liquid (GTL) plant for offshore applications. The coupled cogeneration of NH_3 increases the commercial feasibility of the GTL process and exploits the already available N_2 and H_2. The combined GTL-NH_3 process is self-sufficient as for power and water demands and, thus, suitable for application in remote locations [106].

The naphtha reforming process as the endothermic reaction has been coupled with ammonia synthesis [107]. Increased thermal efficiency and reduced operating costs were reported together with reduced thermal load of condensers for the ammonia loop. The results were compared with conventional reactors, showing an increase in aromatics yield, but reduced conversion of nitrogen. Some parameters were tuned through a genetic algorithm to find the best solution, which was characterised by some penalty for ammonia productivity.

Similarly, the simultaneous production of ammonia and methyl ethyl ketone (MEK) has been investigated with a plant model [108], comparing co-current and counter-current flow configurations. The heat released from the exothermal ammonia synthesis reaction inside heat exchanger reactor tube supplies the required heat for the endothermic 2-butanol dehydrogenation reaction in the co-axial shell. The hydrogen co-produced with MEK by dehydrogenation of 2-butanol is used to supply ca. 30% of the ammonia synthesis feed in the exothermic side. Different conceptual layouts have been also proposed [108, 109].

Furthermore, a solid oxide electrolyser (SOE) can efficiently exploit the reaction enthalpy to produce renewable hydrogen. An integrated ammonia synthesis reactor, a SOE and an air separator have been simulated with Aspen Plus to develop a plant model. A reduction in 40% energy input has been claimed [110].

A production facility has been simulated and economically evaluated to produce ammonia from air and water as raw materials. N_2 is separated from O_2 through the Linde process. H_2 is produced by water splitting. A three-bed packed reactor with either direct or indirect cooling is presented. Duties for compression and electrolysis are derived from renewable sources, such as photovoltaic or wind. Too high investment is envisaged (1.5 GEuro), though promising production cost of ammonia, i.e. 1.35 Euro/kg [111].

6 Conclusions

To conclude, the evolution of ammonia synthesis plants is double, on the one hand super-giant plants, with capacity of 2000 t/d, are under construction, however accompanied by spread medium–small-scale plants (400–600 t/d). The steam reforming of natural gas is the best, most energy-intensive option, provided that this raw material is available. Heavy fuel oil or coal is still valid alternatives for some countries, such as China, due to largely available feedstocks [10]. Radial flow converters with small particle size catalysts and indirect cooling are currently the preferred options.

This process is at the basis of industrial chemistry since a century, but renovation of reactor concepts is continuous to allow process intensification. This is the key to ensure the economic sustainability and availability of cheap raw material for the fertilisers industry.

References

1. https://pubs.er.usgs.gov/publication/70170140 (n.d.)
2. Kandemir T, Schuster ME, Senyshyn A, Behrens M, Schlögl R (2013) Angew Chemie Int Ed 52:12723
3. Cherkasov N, Ibhadon AO, Fitzpatrick P (2015) Chem Eng Process Process Intensif 90:24
4. Lan R, Irvine JTS, Tao S (2011) Int J Hydrogen Energy 37:1482
5. Zhu ZL, Chen DL (2002) Nutr Cycl Agroecosyst 63:117
6. Potter P, Ramankutty N, Bennett EM, Donner SD (2010) Earth Interact 14:1–22
7. Reese M, Marquart C, Malmali M, Wagner K, Buchanan E, McCormick A, Cussler EL (2016) Ind Eng Chem Res 55:3742
8. Brown DE, Edmonds T, Joyner RW, McCarroll JJ, Tennison SR (2014) Catal Lett 144:545
9. Appl M (1999) Ammonia. Principles and industrial practice. Wiley-VCH, Weinheim
10. Jennings J (ed) (1991) Catalytic ammonia synthesis. Fundamentals and practice. Springer Science Business Media LCC
11. Liu H (2013) Ammonia synthesis catalysts. Innovation and practice. Chemical Industry Press; Wold Scientific
12. Slack A, James G (eds) (1974) Ammonia, vol 2. Marcel Dekker, New York
13. Nielsen A (ed) (1995) Ammonia. Catalysis and manufacture. Springer, Berlin
14. Pernicone N, Ferrero F, Rossetti I, Forni L, Canton P, Riello P, Fagherazzi G, Signoretto M, Pinna F (2003) Appl Catal A Gen 251:121
15. Liu H (2014) Cuihua Xuebao/Chinese. J Catal 35:1619
16. Forni L, Pernicone N (2003) PCT Int Appl. WO 2002-EP11707 20021018 (2002)
17. Carpenter D, Maloney K (2012) US9272920B2
18. Lin B, Wei K, Ni J, Lin J (2013) ChemCatChem 5:1941
19. Kowalczyk Z, Jodzis S, Raróg W, Zieliński J, Pielaszek J (1998) Appl Catal A Gen 173:153
20. Forni L, Molinari D, Rossetti I, Pernicone N (1999) Appl Catal A Gen 185:269
21. Kowalczyk Z, Krukowski M, Raróg-Pilecka W, Szmigiel D, Zielinski J (2003) Appl Catal A Gen 248:67
22. Rossetti I, Pernicone N, Forni L (2001) Appl Catal A Gen 208:271
23. Rossetti I, Pernicone N, Forni L (2003) Appl Catal A Gen 248:97
24. Rossetti I, Forni L (2005) Appl Catal A Gen 282:315
25. Rossetti I, Pernicone N, Forni L (2005) Catal Today 102–103:219
26. Rossetti I, Mangiarini F, Forni L (2007) Appl Catal A Gen 323:219

27. Muhler M, Rosowski F, Hinrichsen O, Hornung A, Ertl G (1996) Stud Surf Sci Catal 101:317
28. Mahdi W, Sauerlandt U, Wellenbüscher J, Schütze J, Muhler M, Ertl G, Schlögl R (1992) Catal Lett 14:339
29. Szmigiel D, Bielawa H, Kurtz M, Hinrichsen O, Muhler M, Raróg W, Jodzis S, Kowalczyk Z, Znak L, Zieliński J (2002) J Catal 205:205
30. Gillespie LJ, Beattie JA (1930) Phys Rev 36:743
31. Tripodi A, Compagnoni M, Bahadori E, Rossetti I (2018) J Ind Eng Chem 66:176
32. Temkin MI, Pyzhev V (1940) Acta Physicochim URSS 327
33. Jennings JR (1991) Catalytic ammonia synthesis. Springer, Boston
34. Rossetti I, Pernicone N, Ferrero F, Forni L (2006) Ind Eng Chem Res 45:4150
35. Buzzi Ferraris G, Donati G, Rejna F, Carrà S (1974) Chem Eng Sci 29:1621
36. Yu BY, Chien IL (2015) Ind Eng Chem Res 54:10073
37. Aráujo A, Skogestad S (2008) Comput Chem Eng 32:2920
38. Straus J, Skogestad S (2018) IFAC-PapersOnLine 51:536
39. Straus J, Skogestad S (2017) In: Proceedings of the 2017 21st international conference process control (PC), 309
40. Arora P, Hoadley AFA, Mahajani SM, Ganesh A (2016) Ind Eng Chem Res 55:6422
41. Andersson J, Lundgren J (2014) Appl Energy 130:484
42. Vancini CA (1961) La Sintesi Dell'Ammoniaca. Hoepli, Milan
43. Evans B, Hawkins S, Schulz G (eds) (1991) Ullmann's encyclopedia of industrial chemistry. VCH, Weinheim
44. Simiceanu J, Petrila C, Pop A (1983) Chem Tech 35:628
45. Singh CPP, Saraf DN (1979) Ind Eng Chem Process Des Dev 18:364
46. Gaines LD (1979) Ind Eng Chem Process Des Dev 18:381
47. Lukas DGK (1962) Chem Eng 7:336
48. Li T, Xu MS, Zhu BC, Fang DY, Ying WY (2009) Ind Eng Chem Res 48:8926
49. Yancy-Caballero D, Biegler LT, Guirardello R (2015) Chem Eng Trans 43:1297
50. Yusup S, Zabiri H, Yusoff N, Yew YC (2006) In: Proceedings of the Fifth WSEAS international conference on data networks, communications, computers, p 258
51. Méndez González JM, Díaz de León Cabrero M (2016) Chem Eng Res Des 113:256
52. Nguyen TAN, Nguyen TA, Vu TD, Nguyen KT, Dao TKT, Huynh KPH (2017) In: IOP conference series: materials science and engineering, vol 206
53. Cheema II, Krewer U (2018) RSC Adv 8:34926
54. Angira R (2011) Int J Chem React Eng 9
55. Vorres K (1980) Energy Res 4:109
56. Wurzbacher G (1977) Chern Tech 6:317
57. Teggers H, Jiintgen H (1984) Petrochem Brenns Chern 37:163
58. Staege H (1979) Chern Eng 86:106
59. Michaels H, Leonard H (1978) Chern Eng Prog 74:85
60. Staege H (1982) Hydrocarbon Process 61:92
61. Sharpe R (1976) Hydrocarbon Process 55:171
62. Hiller H (1975) Erdal Kohle. Erdgas, Petrochem Brenns Chern 28:74
63. Konkol W, Ruprecht P, Cornils B, Diirrfeld R, Langhoff J (1982) Hydrocarbon Process 61:97
64. Cornils B, Hibbel J, Ruprecht P, Diirrfeld R, Langhoff J (1981) Hydrocarbon Process 60:149
65. Schlinger W (1980) Energy Res 4:127
66. Ting A, Wan S-W (1969) Chern Eng 76:185
67. Moe J (1962) Chern Eng Prog 58:33
68. Habermehl R (1965) Chern Eng Prog 61:57
69. Buckthorp C (1978) Nitrogen 113:34
70. de Guido G, Compagnoni M, Pellegrini LA, Rossetti I (2018) Front Chem Sci Eng 12:315–325
71. Elberling K, Gabriel W (1977) Chern Tech 29:43
72. Sehrt B, Polster P (1980) Chern Tech 32:345
73. Shah V, McFarland J (1988) Hydrocarbon Process 67:43
74. Shah V (1988) Energy Prog 8:67

75. Volkamer K, Wagner E, Schubert F (1982) Plant/Oper Prog 1:134
76. Royse S (1989) Process Eng 70:36
77. Hein L (1952) Chern Eng Prog 48:412
78. Fischer CD, Iribarren OA (2012) Ind Eng Chem Res 51:16410
79. LeBlanc JJ, Madhavan S, Porter R (1978) In: Kirk-Othmer encyclopedia of chemical technology, vol 2, 3rd edn. Wiley, New York, p 495
80. Quartulli O, Fleming J, Finneran J (1968) Hydrocarbon Process 47:153
81. Nielsen A (1969) Chim Ind 51:1052
82. Hansen H (1968) US 3372988
83. https://image.slidesharecdn.com/fyppresentation-170101163229/95/final-year-project-presentation-11-638.jpg?cb=1483288377 (n.d.)
84. Farivar F, Ebrahim HA (2014) Chem Prod Process Model 9:89
85. Azarhoosh MJ, Farivar F, Ale Ebrahim H (2014) RSC Adv 4:13419
86. Sergeev SP, Krasnushkina NV, Maidurov NP, Petrovskaya MA (2011) Catal Ind 3:283
87. Zardi U, Pagani G (1988) EP 287765
88. Farivar F, Ebrahim HA (2014) RSC Adv 4:48293
89. Zardi U (1982) Hydrocarbon Process 61:129
90. Allen J (1965) Chern Process Eng 46:473
91. Handman S, Leblanc JJ (1983) Chern Eng Prog 79:56
92. Peterson R, Finello R, Denavit G (1985) EP 134260
93. Dashti A, Khorsand K, Marvast MA, Kakavand M (2006) Pet Coal 48:15
94. Khademi MH, Sabbaghi RS (2017) Chem Eng Res Des 128:306
95. Adhi TP, Akbar FI (2018) In: MATEC web of conference, vol 156, p 03047
96. Saviano F, Lagana V, Bisi P (1981) Hydrocarbon Process 60:99
97. Lavie R (1985) Chem Eng Sci 40:2019
98. Liu H, Li X, Hu Z (2001) Chem React Eng Process 17:28
99. Flórez-Orrego D, de Oliveira S Jr (2017) Energy 137:234
100. Ghannadzadeh A, Sadeqzadeh M (2016) Energy Convers Manage 109:63
101. Penkuhn M, Tsatsaronis G (2017) Energy 137:854
102. Chen C, Zhao L, Lavine AS (2018) Sol Energy 176:638
103. Chen C, Zhao L, Kong M, Lavine AS (2018) Sol Energy 176:256
104. Abdiwe R, Haider M (2017) In: AIP conference proceedings, vol 1850
105. Chen C, Lovegrove KM, Sepulveda A, Lavine AS (2018) Sol Energy 159:992
106. Ostadi M, Hillestad M (2018) Chem Eng Technol 41:1668
107. Shakeri M, Iranshahi D, Naderifar A (2019) Chem Eng Process Process Intensif 138:15
108. Ghani R, Iranshahi D (2019) Int J Hydrogen Energy 44:2905
109. Ghani R, Iranshahi D (2019) Appl Therm Eng 154:238
110. Cinti G, Frattini D, Jannelli E, Desideri U, Bidini G (2017) Appl Energy 192:466
111. Sánchez A, Martín M (2018) J Clean Prod 178:325

A Review on Ammonia Derivatives as Corrosion Inhibitors for Metals and Alloys

Chandrabhan Verma, M. A. Quraishi and Eno E. Ebenso

Abstract Nowadays, metallic corrosion is one of the most challenging and damaging phenomena that is associated with huge economic and safety losses. The corrosion damages become more severe during some industrial cleaning processes where metallic materials are allowed for their surface treatment for their further processing and applications. Therefore, corrosion inhibitors, mostly organic compounds are employed during these processes. Most of the investigated inhibitors are associated with salient features, including the presence of heteroatoms (N, S, O, P, etc.) and aliphatic chains and aromatic rings. Generally, organic inhibitors adsorb and form defensive film that isolates the metal(s) from surrounding corrosive environment and protect from corrosion. Literature survey reveals that ammonia derivatives are highly used in almost all kinds of electrolytic media. Aliphatic and aromatic amines, amino acids, hydrazine and phenylhydrazine and their derivatives are most frequently used ammonia derivatives as corrosion inhibitors. Present chapter dealing with the collection of few major reports available on the topic "ammonia derivatives as corrosion inhibitors." Amino acids (AAs) can also be regarded as mono-substituted amines; therefore, present chapter also includes literature available on AAs as corrosion inhibitors.

Keywords Corrosion inhibitors · Mixed type · Ammonia and its derivatives · Langmuir adsorption isotherm · DFT calculations · DM simulations

C. Verma (✉) · E. E. Ebenso
Department of Chemistry, School of Chemical and Physical Sciences and Material Science Innovation & Modelling (MaSIM) Research Focus Area, Faculty of Natural and Agricultural Sciences, North-West University, Private BagX2046, Mmabatho 2735, South Africa
e-mail: chandraverma.rs.apc@itbhu.ac.in; cbverma38@gmail.com

M. A. Quraishi
Center of Research Excellence in Corrosion, Research Institute, King Fahd University of Petroleum and Minerals, Dhahran 31261, Saudi Arabia

Inamuddin et al. (eds.), *Sustainable Ammonia Production*,
Green Energy and Technology, https://doi.org/10.1007/978-3-030-35106-9_3

1 Introduction

Corrosion is defined as the deterioration of metallic materials (metals and alloys) and their properties by their interactions with the constituents of surroundings [1]. Corrosion is a highly challenging and damaging phenomenon that results in very high safety and economic damages. A corrosion cost study conducted by NACE in 2016 revealed that the annual cost of corrosion was $2.5 trillion USD that constituted about 4.2% of world GDP [2]. Corrosion is a universal problem for numerous countries including USA, Australia, Japan, China, Korea, India and South Africa, etc. In India and South Africa, annual corrosion cost is about $100-billion USD and $9.6-billion USD, respectively [3]. Nowadays, metallic corrosion and its inhibition are attracting the particular attention of scientists and engineers because of increased use of metallic materials in household and industrial applications. Because of their very high mechanical supremacy and profitable behavior, metallic materials are frequently used as constructional materials for various industries, including oil-gas and petroleum industries, bridges, railway tracks and transportation pipelines. Corrosion inhibition of these metallic assets requires very high maintenance, economic and safety costs [4]. Apart from the corrosion of constructional resources dissolution of metallic components, cleaning processes are also connected with immense economic losses. In order to avoid the unwanted metallic loss during aggressive cleaning processes, inhibitors are added in the aggressive solution. Several types of corrosion monitoring techniques have been employed previously in order to reduce or avoid the unwanted loss of metallic materials via corrosion. Figure 1 represents some practices of metallic corrosion inhibition. Most of the inhibitors contain electron-rich substituents such as $>C=O$, $>C=N$, $-N=O$, $>C=S$, $-N=N-$, $-COO-R$, $-CO-NH_2$, $-CO-X$, $-C\equiv N$, $-C\equiv C-$, etc., and multiple bonds of functional groups and aromatic

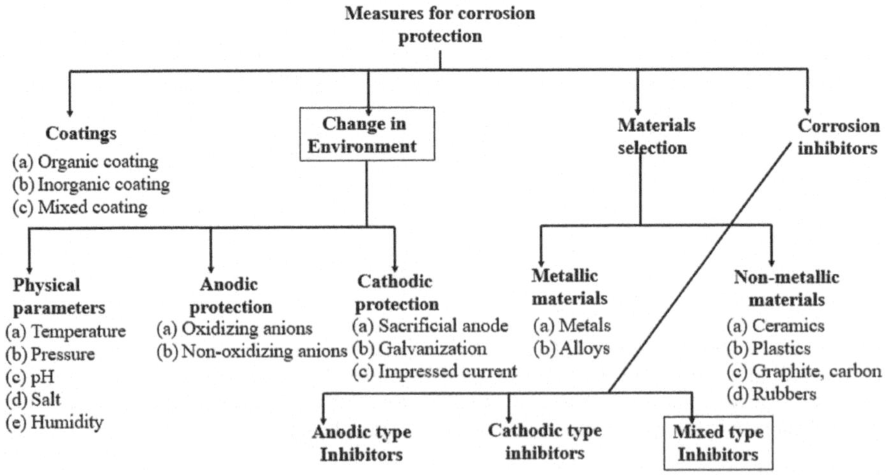

Fig. 1 Measures for corrosion protection

rings that help in their adsorption on metallic surface [5, 6]. Mechanism of metallic corrosion in the absence of organic inhibitors can be accessible as follows [7, 8]:

$$M + nH_2O \longleftrightarrow M(H_2O)n_{ads} \quad (1)$$

$$M(H_2O)n_{ads} + 2X^- \longleftrightarrow M[(H_2O)_n(X)_2^-]_{ads} \quad (2)$$

$$M[(H_2O)_n(X)_2^-]_{ads} \rightarrow M[(H_2O)_n(X)_2]_{ads} + 2e^- \quad (3)$$

$$M[(H_2O)_n(X)]_{ads} \rightarrow M^{2+} + OH^- + 2X^- + H^+ \quad (4)$$

Whereas, in the presence of inhibitors, the following mechanism can be operated [7, 8]:

$$M + nH_2O \longleftrightarrow M(H_2O)n_{ads} \quad (5)$$

$$M(H_2O)n_{ads} + 2X^- \longleftrightarrow M[(H_2O)_n(X)_2^-]_{ads} \quad (6)$$

$$M(H_2O)_n(X_2^-)]_{ads} + INH - H^+ \rightarrow M(H_2O)_n(X)_2INH - H]_{ads}^- \quad (7)$$

In the above equations, M represents the metal; X is the counter ion of the aggressive medium, e.g., $C\bar{l}$ of HCl and INH is the inhibitor. In acidic medium rapid oxidation of metal outcomes addicted to the gathering of the positive charges (M^{n+}) on the surface, which attracts the positively charged counter ions (X) by electrostatic force of attraction.

Therefore, firstly organic inhibitors adsorb by using their electrostatic force of attraction. However, after the release of hydrogen at cathode inhibitors become neutral with the free unshared pair of electrons on heteroatoms. The unshared electron pairs are transferred into the vacant d-orbital that results in the co-ordination bonding (chemisorptions) between the inhibitors and the metallic surface. This is recognized as a donation. However, in order to minimize inter-electronic repulsion, some of the metallic electrons are transferred to empty anti-bonding orbitals through retro-donation. Schematic illustration of the binding between metal and inhibitors is presented in Fig. 2.

2 Ammonia Derivatives as Corrosion Inhibitors

Ammonia and its derivatives including hydrazine and aliphatic and aromatic amines are used as corrosion inhibitors for a variety of conditions. Various nitrogenous heterocyclic compounds, including pyridine, imidazole, pyrrole, furan and thiophene, purine and pyrimidine and pyrazine derivatives are employed as effective inhibitors

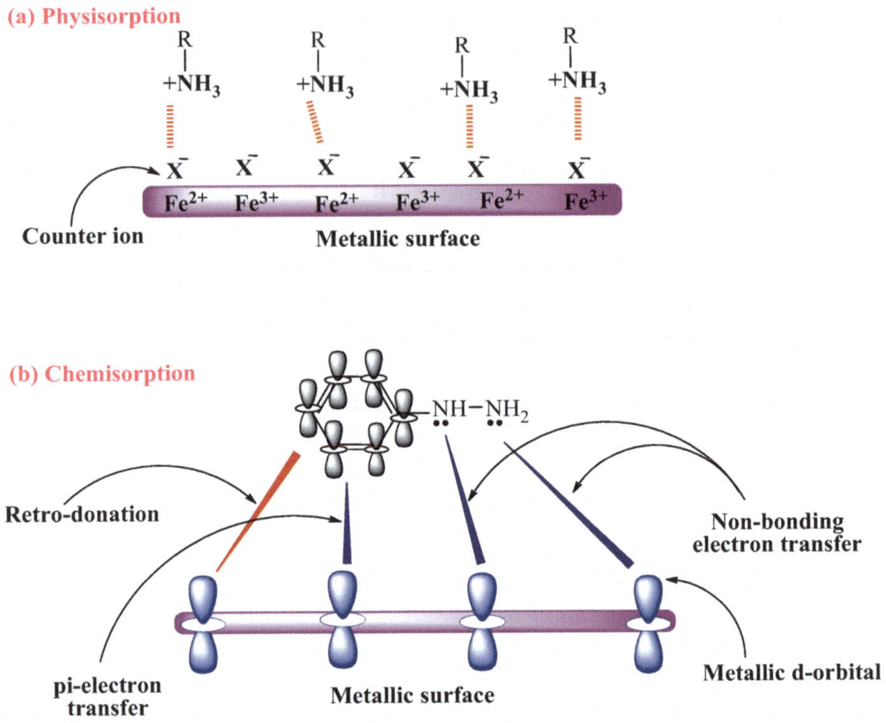

Fig. 2 Mechanism of physisorption (**a**) and chemisorption (**b**) of organic nitrogenous inhibitors in acidic medium

[9]. Inhibitiveness of these inhibitors depends upon abundant aspects, for example, solubility, temperature, pressure, presence of humidity and salts and their structure. Generally, an inhibitor having higher solubility shows better inhibition efficiency than the inhibitor having lower solubility. Increase in temperature causes a significant reduction in the protection efficiency of the inhibitor molecules which can be attributed to the fragmentation, rearrangement and/or decomposition of the inhibitor molecules at the high temperature [9, 10]. More so, increased kinetic energy at elevated temperature can also decrease the attraction between inhibitor molecules and metallic surface. Apart from these, electronic structure of the inhibitor molecules has marked effect on their protectiveness as an inhibitor partaking electron-donating substituents for instance –NH_2, –NHMe, –NMe_2, –OH, –OCH_3, etc., shows better protection efficiency than their counter parts such as –NO_2, –CN and –COOEt. Effect of substituents can be described using Hammett equations accessible below [9, 10]:

$$\log = \frac{K_R}{K_H} = \rho\sigma \tag{8}$$

$$\log = \frac{1 - \eta\%_R}{1 - \eta\%_H} = \rho\sigma \tag{9}$$

$$\log = \frac{\eta\%_R}{\eta\%_H} = \log\frac{C_{rH}}{C_{rR}} = \rho\sigma - \log\frac{\theta_R}{\theta_H} \quad (10)$$

In the above Hammett equations, σ represents the value of Hammett constant that gives information about the electronic influence of the substituents on the center of metal-inhibitor interactions. Generally, its positive value indicates the electron-withdrawing effect of the substituents, whereas the converse is true for substituents having negative values of Hammett constant. The K, $\eta\%$ and θ represent the equilibrium constant for metal-inhibitor reactions, inhibition efficiency and surface coverage for inhibitors without any substituent (−H) and with substituent (−R). Obviously, the negative value of Hammett constant is interrelated with electron-donating aptitude, and therefore, high protection efficiency of the inhibitor molecules and the converse is true for the positive value of Hammett constant. Present chapter mainly focuses on the collection of corrosion inhibition effect of ammonia derivatives, including aliphatic and aromatic amines and hydrazine derivatives.

The aliphatic and aromatic amines are highly used because of their high basicity and high electron-donating ability that generally offers strong metal-inhibitor interactions. Amines represent the open chain or cyclic aliphatic class of mono-, di- or tri-substituted ammonia derivatives. Generally, in amines polar head containing nitrogen (amine group) interact/attach with metallic surface and non-polar hydrophobic tail flag into the electrolyte. Therefore, amines form surface protective films avoid the contact of metallic surface with corrosive solution. Inhibitiveness of the amines is fundamentally contingent upon the nature and length of the hydrophobic carbon chain. Obviously, high inhibition effectiveness is associated with larger hydrophobic carbon chain. Several amines are investigated (as inhibitors) in numerous corrosive media, including HCl, NaCl and H_2SO_4. Inhibition property of two amines namely, 1,8-diaminooctane and tetraethylenepentamine and their derivatives were evaluated for their anticorrosive behavior on carbon steel [11]. Investigated amine derivatives performed as mixed type inhibitors and their efficiency was concentration dependent. Adsorption of the amine derivatives obeyed the Langmuir adsorption isotherm. The inhibition effect of N-heterocyclic compounds on iron corrosion in 1 M $HClO_4$ showed that they are acted as mixed type inhibitors and their adsorption obeyed the Langmuir adsorption isotherm [12]. It was further observed that inhibition effect of the investigated amines further depends upon the electron density over the nitrogen atom. Yadav and associates [13] described the inhibition consequence of two substituted amines on N80 steel in 15% HCl solution using. Both the amine derivatives acted mixed type inhibitors. In another study [14], the inhibition effect of amine-based surfactants on carbon steel corrosion in formation water (deep well water) was examined. Investigated amine-based inhibitors showed 86–96% inhibition efficiencies at different concentrations ranging from 200 to 400 ppm. Results showed that inhibition efficiency upsurges on enhancing the length of the hydrophobic chain and the maximum value of 96% were obtained for surfactant IV. They behaved as anodic type inhibitors and their adsorption obeyed the Langmuir adsorption isotherm. The effect of two amines on carbon steel corrosion in seawater system has been investigated elsewhere [15]. Further, the effect of hydrophobic chain length on other amine derivatives has also been investigated in literature [16–18]. Generally, the inhibition efficiency of the amines derivatives increases on increasing the length of hydrophobic chain; however, increase in the chain of carbon chain beyond certain limit decreases the inhibition efficiency. This is because increase in hydrophobic chain length causes significant decrease in the solubility of the inhibitors into the polar media and therefore inhibition efficiency. Beside mild and carbon steel, amines by-products have also been used as inhibitors for other metals [19–22]. Some common representatives are presented in Table 1.

Aniline (also known as benzenamine or aminobenzene) and its derivatives represent a special class of biologically and industrially useful compounds including in the field of metallic corrosion inhibition [23]. Aniline can be regarded as one of the most important aromatic derivatives of ammonia which serves as starting chemical for the synthesis of several industrially and biologically useful nitrogenous compounds [24]. Generally, they are synthesized by the reduction of their nitro-analog. Aniline and its derivatives are associated with high electron density that resulted due to availability of non-bonding electrons of nitrogen and π-electrons of the aromatic ring therefore generally they act as efficient corrosion inhibitions [25]. In view of this, pyridine and its numerous derivatives derived from natural and synthetic sources have been employed as corrosion inhibitors. Henrıquez-Roman and coworkers [26] studied the inhibition effect of aniline and its para-substituted methyl, methoxy and nitro derivatives on corrosion of copper in 1 M HCl solution using computational studies. They demonstrated that investigated inhibitors preferably adsorb on the active places present over the metallic surface. Nature of substituent greatly affects the adsorption tendency of the inhibitor molecules. Similar observation was reported by another group of authors [27] while demonstrating the effect of aniline and p-bromoaniline, m-chloroniline, o-chloroaniline, p-chloroaniline and p-methyl aniline on corrosion of copper in phosphoric acid solution having different concentrations. Results showed that methyl aniline and aniline showed the highest and lowest protection efficiencies, respectively. Interactions of the investigated inhibitor molecules obeyed the Flory-Huggins and Langmuir adsorption isotherm models. Inhibition property of the azo rhodanine derivatives of aniline for protection copper corrosion in acidic medium has also been studied elsewhere [28]. Nasser [29] functionalized brominated Jojoba oil using aniline and its two derivatives and tested them for their anticorrosive property on mild steel in 1 M HCl. They observed that inhibitor II which is constituted by 2, 4-dimethyl aniline displayed the uppermost protection efficiency among the investigated inhibitors. Inhibition effect of numerous anilines along with pyridine derivatives on steel corrosion in 1 M HCl was reported elsewhere [30]. In the study basicity values of the investigated compounds were correlated with their protection efficiency,

Table 1 Chemical structure, nature of metals and electrolyte and other relevant information about aliphatic and aromatic amines and hydrazine derivatives as corrosion inhibitors

S. No.	Chemical structure	Nature of inhibitors	Nature of metal and electrolyte	Techniques	Salient features	References
1	R–NH_2 R = $-C_6H_{13}$, $-C_8H_{17}$, $-C_{10}H_{21}$, $-C_{12}H_{25}$	Mixed type, Frumkin adsorption isotherm	Mild steel/2M HCl	WL and PDP	Effect of hydrophobic substituents was evaluated on corrosion inhibition on amines. inhibition efficiencies followed the order: $-C_{12}H_{25} > -C_{10}H_{21} > C_8H_{17} > C_6H_{13}$	[16]
2	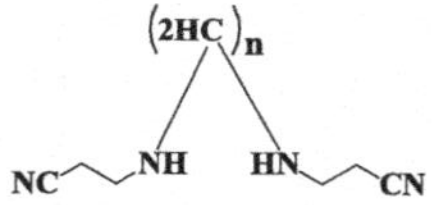 n = 2, 3, 6	Mixed type, Langmuir adsorption isotherm	Mild steel/1 M HCl	WL, PDP and DFT	Inhibition efficiency was evaluated as different temperature and it was observed that hydrophobic chains exert significant effect on the inhibition effect of investigated amines	[17]
3	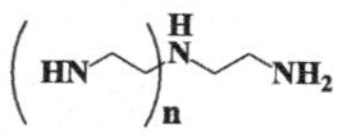 n = 0, 1, 3	Mixed type inhibitors	Fe/sea water	EIS, PDP, SEM and EDX	Inhibition efficiency increases with increasing the carbon chain length (hydrophobicity) of the carbon chain	[18]

(continued)

Table 1 (continued)

S. No.	Chemical structure	Nature of inhibitors	Nature of metal and electrolyte	Techniques	Salient features	References
4	$H_2N{-}(CH_2)_n{-}NH_2$ n = 0, 2, 6, 8	Mixed type/predominantly cathodic	Copper/0.1 M H_2SO_4	EIS, PDP and DFT	Inhibition efficiency largely influence by the value of n. Generally, greater the value of n greater was efficiency	[18]
5	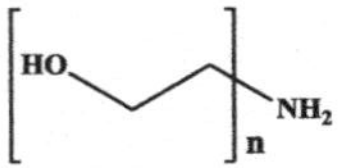 n = 0, 1,2	Mixed type/Freundlich adsorption	Al/1 M H_3PO_4	WL, PDP and DFT	Inhibition efficiency increases on increasing the number of hydroxyethyl moieties, i.e., order of inhibition efficiencies are consistent as: n = 3 > n = 1 > n = 0	[19]
6	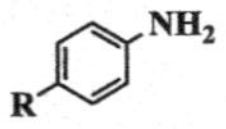 R = –H, –Cl, CH3, –OCH3, –NO2	Eads values were 30–67 kcal/mol	Copper/1 M HCl	Computational studies	Results showed that inhibitor molecules preferably adsorb on the more active sites in which substituents exert significant effect	[26]

(continued)

Table 1 (continued)

S. No.	Chemical structure	Nature of inhibitors	Nature of metal and electrolyte	Techniques	Salient features	References
7	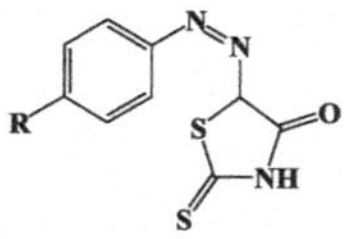 R = –OCH_3 (I), –H (II) and –NO_2 (III)	Mixed type, Langmuir adsorption isotherm	Copper/2 M HCl	WL, EIS, PDP, SEM, EFM and DFT	Presence of nitro group decreases the inhibition efficiency, whereas the presence of methoxy group increases efficiency as compared to non-substituted compound	[28]
8	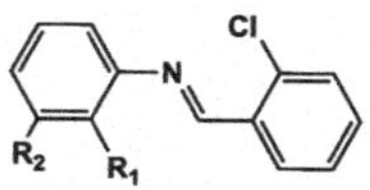 NCCM: R_1= -CH_3; R_2=-Cl NCF: R_1= -F; R_2=-H	Mixed type, Langmuir adsorption isotherm	Carbon steel/1 M HCl	EIS, PDP, SEM, DFT and MDs	Among the tested inhibitors, NCF showed better protection as compared to the NCCM. Experimental results are supported by DFT and MDS studies	[32]

(continued)

Table 1 (continued)

S. No.	Chemical structure	Nature of inhibitors	Nature of metal and electrolyte	Techniques	Salient features	References
9	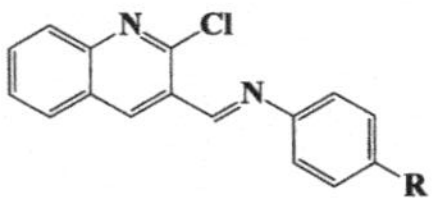 CAP: R = –OH; CMPA: R = –OCH_3, CNPA: R = –NO_2	Cathodic type, Langmuir adsorption isotherm	Mild steel/1 M HCl	WL, EIS, PDP and SEM	Substituent effect was observed in the study. Effectiveness of the inhibitors followed the order: CAP > CMPA > CNPA	[33]
10	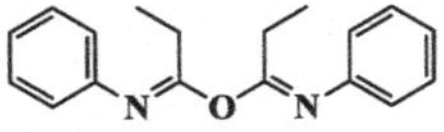 PPB	Mixed type, Langmuir adsorption isotherm	Carbon steel/1 M HCl	EIS, PDP and DFT	PPB acted as efficient corrosion inhibitor and it adsorbed spontaneously over the metallic surface	[34]
11	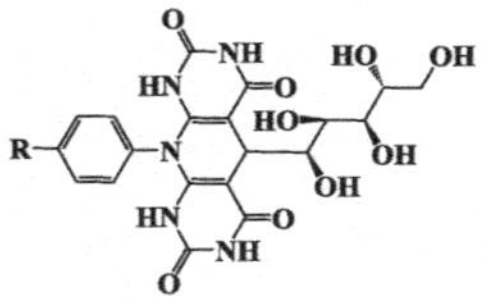	Mixed type, Langmuir adsorption isotherm	Mild steel/1 M HCl	WL, EIS, PDP, SEM, AFM, DFT and MCS	Methoxy (–OCH3) substituted GPH molecule showed the highest protection efficiency of 97.82% at 10.15×10^{-5} mol concentration	[23]

(continued)

Table 1 (continued)

S. No.	Chemical structure	Nature of inhibitors	Nature of metal and electrolyte	Techniques	Salient features	References
12	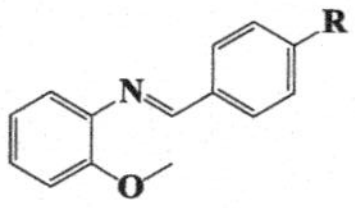 (a)R = –H, (b)R = $-CH_3$, (c)R = –Cl, (d)R = $-NO_2$	Mixed type, Langmuir adsorption isotherm	Aluminum/1 M HCl	EIS, WL and DT	EIS and weight loss methods were used for the evaluation of inhibition effect of inhibitors. Experimental results were supported by DFT study. Inhibition efficiencies followed the order: –H> $-CH_3$> –Cl> $-NO_2$	[35]

and it was observed that corrosion rate was lesser for the inhibitor having high basicity. In another study [31], it is demonstrated that p-methoxy aniline showed better protection efficiency followed by p-methyl aniline and finally by aniline onQD36 steel corrosion in 0.5 H HCl. Inhibitive effect of aniline and its derivatives are also reported elsewhere [23, 32–34]. Anticorrosive behavior of D-glucose-based heterocyclics for mild steel in 1 M HCl showed that inhibition effectiveness followed the order: GPH-3 (–OCH_3)> –GPH-2 (–OH) > GPH (–OH) [23]. Table 1 represents some main collection on aniline-based corrosion inhibitors. Aniline-based corrosion inhibitors are also tested for aluminum [35] and zinc [36–38].

Hydrazine derivatives represent a special class of ammonia derivatives in which two nitrogen joined together by nitrogen-nitrogen covalent bond. Hydrazine and its derivatives (mostly phenyl hydrazine) are used in versatile condition. Generally, their extraordinary inhibitiveness is resulted due to the availability of unshared electron pairs of nitrogen and widespread conjugation responsible or strong metal-inhibitor interactions. Hydrazine derivatives are tested as inhibitors in another study [39]. A good agreement in the results of DFT and experimental studies was observed. Three similar hydrazine derivatives are studied as inhibitors [40]. Chafai [41] synthesized, characterized and investigated the protection efficiency of p-methyl phenyl hydrazine derivative of m-nitro benzaldehyde (E-NBPTH) on carbon steel in 0.5 M H_2SO_4. Four phenyl hydrazones differing in the nature of substituents were synthesized and investigated as effective corrosion inhibitors for brass in hydrochloric acid (2 M) solution [42]. Synergistic consequence of KI on the inhibitiveness of hydrazones was further studies and it was observed that addition of KI in small amount significantly enhanced the inhibition effect of investigated hydrazones.

Carbazides, semicarbazides and thiosemicarbazides are other very important classes of ammonia derivatives that show very high industrial and biological applications. They are also used as corrosion inhibitors and they showed high inhibition effectiveness [43–45]. Among the carbazides, semicarbazides and thiosemicarbazides, thiosemicarbazides have been used most extensively. Some major collection on thiosemicarbazides as corrosion inhibitors are presented in Table 2. Mahgoub and S. M. Al-Rashdi [46] investigated the inhibition effect thiosemicarbazide (TSC) on mild steel in 0.5 M H2SO4 using electrochemical and SEM techniques. Study further suggested that TSC adsorb on the metallic surface using chemisorption mechanism. In another study, the inhibition effect of thiosemicarbazide (TSC) and phenyl thiosemicarbazide (PTSC) on carbon steel corrosion in 2 M H3PO4 solution is reported at their different concentrations using electrochemical techniques [45]. Ameer et al. [47] studied the inhibition effect of numerous thiosemicarbazide derivatives on carbon steel corrosion in phosphoric acid medium using weight loss and electrochemical techniques. Adsorption of the thiosemicarbazide derivatives followed the Flori-Huggins isotherm model and the tested inhibitors acted as mixed type corrosion inhibitors. Inhibition effect of thiosemicarbazide derivatives has also been reported elsewhere for carbon and mild steel and copper [48–55]. Along with thiosemicarbazides, several carbazide derivatives are also studied as corrosion inhibitors [56–59].

Table 2 Chemical structure, nature of metals and electrolyte and other relevant information about thiosemicarbazide derivatives as corrosion inhibitors

S. No.	Chemical structure	Nature of inhibitor (s)	Nature of metal and electrolyte	Techniques	Salient features	References
1	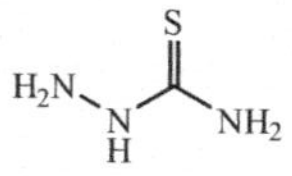 TSC	Mixed type, Langmuir adsorption isotherm	Mild steel in 0.5 M H_2SO_4	PDP, EIS and SEM	Adsorption of the TSC on metallic surface obeyed the chemisorption mechanism and forms the stable inhibitive surface film	[46]
2	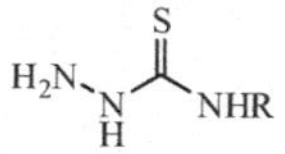 TSC (A): R=-H PTSC (B): R=-Ph	Mixed type, Langmuir adsorption isotherm	Carbon steel/2 M H_3PO_4	PDP and EIS	Inhibition efficiency of both inhibitors increases with their concentration and decreases with temperature	[59]
3	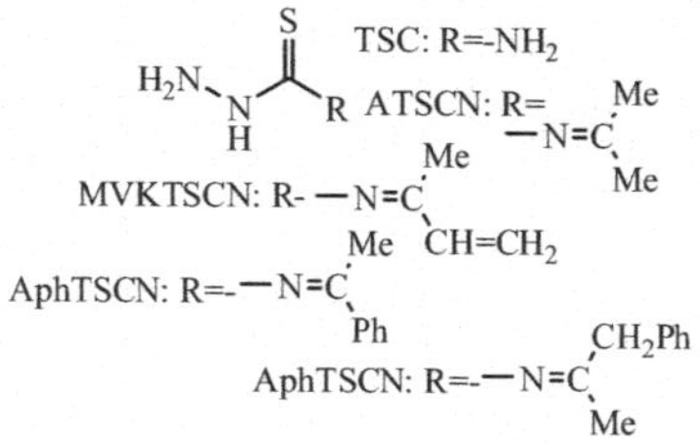	Mixed type, Flory-Huggins adsorption isotherm	Carbon steel/H_3PO_4	WI, PDP and EIS.	Kinetic-thermodynamic model and Flory-Huggins adsorption isotherm model were tested to describe the adsorption of inhibitors on the surface	[47]

(continued)

Table 2 (continued)

S. No.	Chemical structure	Nature of inhibitor (s)	Nature of metal and electrolyte	Techniques	Salient features	References
4	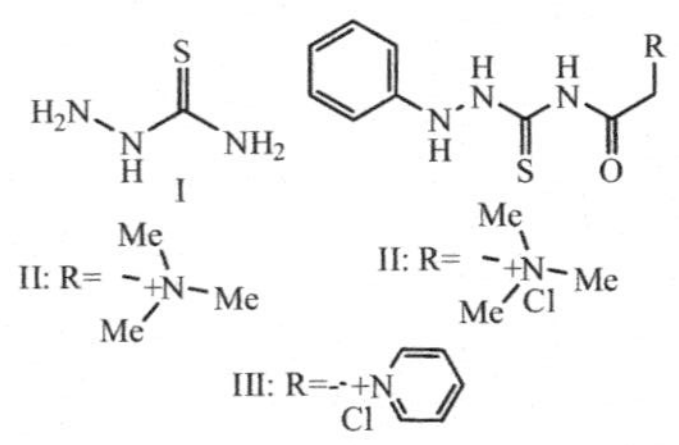	Cathodic type, Temkin adsorption isotherm	Carbon steel/2 M HCl	WL and PDP	Both Temkin adsorption isotherm and kinetic-thermodynamic models were tested to fit the adsorption of inhibitors on the surface. Inhibition efficiency followed the order: I < II < III	[48]
5	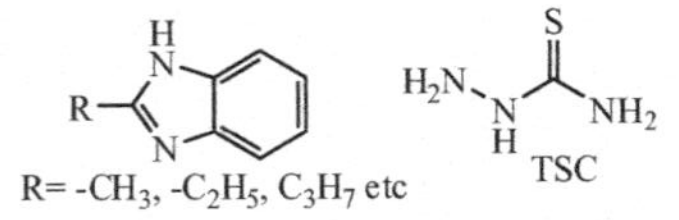	Mixed type, Langmuir adsorption isotherm	Mild steel/M HCl	WL, EIS, PDP, SEM and DFT	Effect of TSC in different benzimidazoles was tested and good agreement in the results of experimental and DFT studies were observed	[49]
6	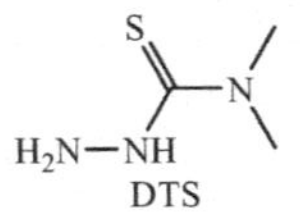	Mixed type, Langmuir adsorption isotherm	Mild steel/2.5 M H_2SO_4	PDP, EIS and DFT	DFT study provides good insight about the inhibition mechanism of DTS	[50]

(continued)

Table 2 (continued)

S. No.	Chemical structure	Nature of inhibitor (s)	Nature of metal and electrolyte	Techniques	Salient features	References
7	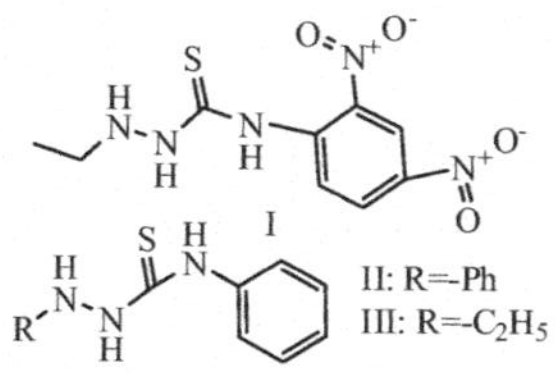	Mixed type, Temkin adsorption isotherm	Carbon steel/2 M HCl	WL, PDP, EIS and DFT	Inhibition efficiency obeyed the order: I > II > III >. Inhibition efficiency increases with the concentration of inhibitors. DFT study provides good support the experimental results	[51]
8	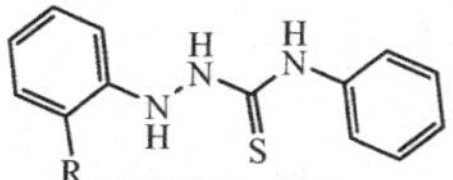 HP4PT: R=-OH D4PT: R=-H AP4PT: R=-NH_2	Langmuir adsorption isotherm	Mild steel/1 M H2SO4	WL, gasometric, DFT and QSAR	Inhibition efficiency of investigated inhibitors followed the order: AP4PT > HP4PT > D4PT	[52]
9	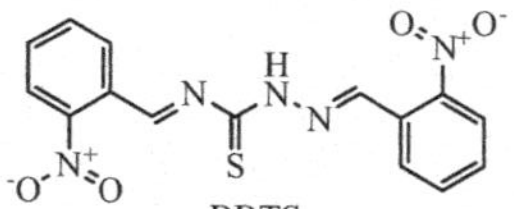 BBTS	Mixed type, Langmuir adsorption isotherm	Mild steel/1 M HCl	EIS, PDP and SEM	BBTS acted as good corrosion inhibitor and its efficiency increases with concentration	[53]

(continued)

Table 2 (continued)

S. No.	Chemical structure	Nature of inhibitor (s)	Nature of metal and electrolyte	Techniques	Salient features	References
10	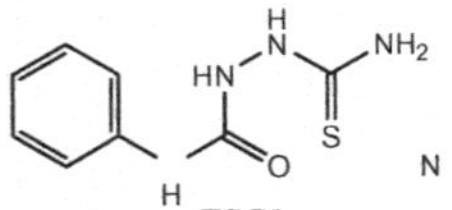 TSC1 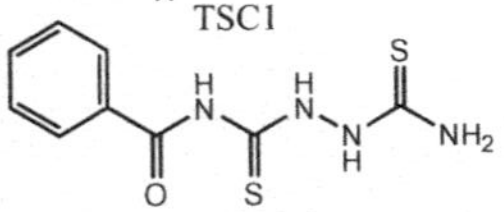 TSC 2	Mixed type, Langmuir adsorption isotherm	Carbon steel/1 M HCl	WL, PDP, EIS, SEM and DFT	Both experimental and computational results showed that TSC 2 showed better inhibition efficiency as compared to the TSC1 and their inhibition efficiencies are concentration and temperature dependent	[54]
11	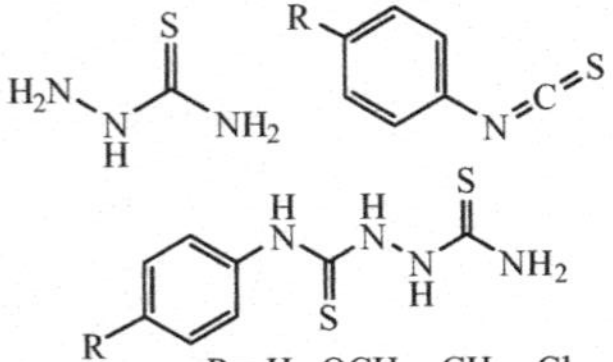 R=-H, $-OCH_3$, $-CH_3$, -Cl	Mixed type adsorption	Copper/NaCl	DFT analysis	DFT study was undertaken in order to support the order of inhibition efficiency of several previously reported inhibitor molecules. Good agreement was observed in two studies	[55]

Acknowledgements Dr. Chandrabhan Verma, thankfully acknowledges the North-West University South Africa for providing financial support under post doctor scheme.

References

1. Mansor N, Abdullah S, Ariffin A, Syarif J (2014) A review of the fatigue failure mechanism of metallic materials under a corroded environment. Eng Fail Anal 42:353–365
2. Gupta RK, Malviya M, Verma C, Quraishi M (2017) Aminoazobenzene and diaminoazobenzene functionalized graphene oxides as novel class of corrosion inhibitors for mild steel: experimental and DFT studies. Mater Chem Phys 198:360–373
3. Verma C, Ebenso EE, Quraishi M (2017) Ionic liquids as green and sustainable corrosion inhibitors for metals and alloys: an overview. J Mol Liq 233:403–414
4. Zaferani SH, Sharifi M, Zaarei D, Shishesaz MR (2013) Application of eco-friendly products as corrosion inhibitors for metals in acid pickling processes–a review. J Environ Chem Eng 1:652–657
5. Emregül KC, Akay AA, Atakol O (2005) The corrosion inhibition of steel with Schiff base compounds in 2 M HCl. Mater Chem Phys 93:325–329
6. Verma C, Olasunkanmi LO, Ebenso EE, Quraishi MA, Obot IB (2016) Adsorption behavior of glucosamine-based, pyrimidine-fused heterocycles as green corrosion inhibitors for mild steel: experimental and theoretical studies. J Phys Chem C 120:11598–11611
7. de Chialvo MG, Chialvo A (1998) Kinetics of hydrogen evolution reaction with Frumkin adsorption: re-examination of the Volmer-Heyrovsky and Volmer-Tafel routes. Electrochim Acta 44:841–851
8. de Chialvo MG, Chialvo A (1999) The Tafel-Heyrovsky route in the kinetic mechanism of the hydrogen evolution reaction. Electrochem Commun 1:379–382
9. Verma C, Olasunkanmi L, Ebenso EE, Quraishi M (2018) Substituents effect on corrosion inhibition performance of organic compounds in aggressive ionic solutions: a review. J Mol Liq 251:100–118
10. Sastri VS (2012) Green corrosion inhibitors: theory and practice. John Wiley & Sons
11. Al-Sabagh AM, Nasser NM, Farag AA, Migahed MA, Eissa AM, Mahmoud T (2013) Structure effect of some amine derivatives on corrosion inhibition efficiency for carbon steel in acidic media using electrochemical and quantum theory methods. Egypt J Pet 22:101–116
12. Babić-Samardžija K, Khaled K, Hackerman N (2005) N-heterocyclic amines and derivatives as corrosion inhibitors for iron in perchloric acid. Anti-Corros Methods Mater 52:11–21
13. Yadav M, Kumar S, Sharma U, Yadav P (2013) Substituted amines as corrosion inhibitors for N80 steel in 15% HCl. J Mater Environ Sci 4:691–700
14. Alsabagh A, Migahed M, Awad HS (2006) Reactivity of polyester aliphatic amine surfactants as corrosion inhibitors for carbon steel in formation water (deep well water). Corros Sci 48:813–828
15. Rihan R, Shawabkeh R, Al-Bakr N (2014) The effect of two amine-based corrosion inhibitors in improving the corrosion resistance of carbon steel in sea water. J Mater Eng Perform 23:693–699
16. De Damborenea J, Bastidas J, Vazquez A (1997) Adsorption and inhibitive properties of four primary aliphatic amines on mild steel in 2 M hydrochloric acid. Electrochim Acta 42:455–459
17. Herrag L, Bouklah M, Patel N, Mistry B, Hammouti B, Elkadiri S, Bouachrine M (2012) Experimental and theoretical study for corrosion inhibition of mild steel 1 M HCl solution by some new diaminopropanenitrile compounds. Res Chem Intermed 38:1669–1690
18. Migahed M, Attia A, Habib R (2015) Study on the efficiency of some amine derivatives as corrosion and scale inhibitors in cooling water systems. RSC Adv 5:57254–57262

19. Fouda A, Abdallah M, Ahmed I, Eissa M (2012) Corrosion inhibition of aluminum in 1 M H_3PO_4 solutions by ethanolamines. Arab J Chem 5:297–307
20. Mahida M, Chaudhari H (2012) Aromatic amines as corrosion inhibitors for zinc in hydrochloric acid. J Chem Pharm Res 4:5195–5201
21. Topal E, Gece G (2017) Untangling the inhibition effects of aliphatic amines on silver corrosion: a computational study. Chem J Moldova 12:64–70
22. El Wanees SA, El Aal Mohamed AA, El Azeem MA, El Said R (2010) Inhibition of silver corrosion in nitric acid by some aliphatic amines. J Dispers Sci Technol 31(11):1516–1525
23. Verma C, Quraishi M, Kluza K, Makowska-Janusik M, Olasunkanmi LO, Ebenso EE (2017) Corrosion inhibition of mild steel in 1 M HCl by D-glucose derivatives of dihydropyrido [2, 3-d: 6, 5-d′] dipyrimidine-2, 4, 6, 8 (1H, 3H, 5H, 7H)-tetraone. Sci Rep 7:44432
24. Antonijevic M, Petrovic M (2008) Copper corrosion inhibitors. A review, Int J Electrochem Sci 3(2008):1–28
25. Goyal M, Kumar S, Bahadur I, Verma C, Ebenso EE (2018) Organic corrosion inhibitors for industrial cleaning of ferrous and non-ferrous metals in acidic solutions: a review. J Mol Liq 256:565–573
26. Henríquez-Román JH, Padilla-Campos L, Páez MA, Zagal JH, Rubio MA, Rangel CM, Costamagna J, Cárdenas-Jirón G (2005) The influence of aniline and its derivatives on the corrosion behaviour of copper in acid solution: a theoretical approach. J Mol Struct (Thoechem) 757:1–7
27. Ayobe EA, Abaza SF, Seleim SM, Ahmed AMM (2012) Anodic corrosion of copper in presence of aniline derivatives
28. Eldesoky A, El-Bindary M, El-Sonbati A, Morgan SM New eco-friendly corrosion inhibitors based on azo rhodanine derivatives for protection copper corrosion
29. Nasser R (2017) Investigation of the behavior of aminated jojoba derivatives as green corrosion inhibitors for mild steel at 0.5 N HCl. Der Chem Sinica 8(2017):123–132
30. Donya A, Pakter M, Shalimova M, Lambin V (2002) The effect of polar substituents in aniline and pyridine derivatives on the inhibition of steel corrosion in acids. Prot Met 38:216–219
31. El-Haddad MN, Fouda AE-AS (2013) Corrosion inhibition effect and adsorption of aniline derivatives on QD36 steel surface in acidic solution. Protection of Metals and Phys Chem Surfaces 49:753–762
32. Belghiti M, Bouazama S, Echihi S, Mahsoune A, Elmelouky A, Dafali A, Emran K, Hammouti B, Tabyaoui M (2017) Understanding the adsorption of newly benzylidene-aniline derivatives as a corrosion inhibitor for carbon steel in hydrochloric acid solution: experimental, DFT and molecular dynamic simulation studies. Arab J Chem
33. Prabhu R, Venkatesha T, Shanbhag A, Kulkarni G, Kalkhambkar R (2008) Inhibition effects of some Schiff's bases on the corrosion of mild steel in hydrochloric acid solution. Corros Sci 50:3356–3362
34. Hayaoui M, Drissi M, Fahim M, Salim R, Rais Z, Mouffarih S, Baba MF, El Hajjaji F, Zarrouk A, Taleb M (2017) Benzenamine derivative as corrosion inhibitor of carbon steel in hydrochloric acid solution: electrochemical and theoretical studies. J Mater Environ Sci 8:1877
35. Ashassi-Sorkhabi H, Shabani B, Aligholipour B, Seifzadeh D (2006) The effect of some Schiff bases on the corrosion of aluminum in hydrochloric acid solution. Appl Surf Sci 252:4039–4047
36. Kumar S, Ladha D, Shah N Adsorption behaviour, thermodynamic and electrochemical studies of chloro derivatives of N-(4-methoxybezylidene) aniline as corrosion inhibitors for zinc in hydrochloric acid
37. Kumar S, Ladha D, Jha P, Shah N (2013) Theoretical study of chloro-N-(4-methoxybenzylidene) aniline derivatives as corrosion inhibitors for zinc in hydrochloric acid. Int J Corros 2013
38. Vashi R, Desai K Aniline as corrosion inhibitor for zinc in hydrochloric acid
39. El Azzouzi M, Aouniti A, Tighadouin S, Elmsellem H, Radi S, Hammouti B, El Assyry A, Bentiss F, Zarrouk A (2016) Some hydrazine derivatives as corrosion inhibitors for mild steel in 1.0 M HCl: weight loss, electrochemical, SEM and theoretical studies. J Mol Liq 221:633–641
40. Belghiti M, Tighadouini S, Karzazi Y, Dafali A, Hammouti B, Radi S, Solmaz R (2016) New hydrazine derivatives as corrosion inhibitors for mild steel protection in phosphoric acid medium. Part A: Exp Study, J Mater Environ Sci 7(2007):337–346

41. Chafai N, Chafaa S, Benbouguerra K, Hellal A, Mehri M (2019) Synthesis, spectral analysis, anti-corrosive activity and theoretical study of an aromatic hydrazone derivative. J Mol Struct 1181:83–92
42. Abdallah M, Al-Agez M, Fouda A (2009) Phenylhydrazone derivatives as corrosion inhibitors for-α-brass in hydrochloric acid solutions. Int J Electrochem Sci 4:336–352
43. Quraishi MA, Jamal D, Singh RN (2001) Inhibition of mild steel corrosion in the presence of fatty acid thiosemicarbazides. Corrosion 28:201–207
44. Kandemirli F, Sagdinc S (2007) Theoretical study of corrosion inhibition of amides and thiosemicarbazones. Corros Sci 49:2118–2130
45. Goulart CM, Esteves-Souza A, Martinez-Huitle CA, Rodrigues CJF, Maciel MAM, Echevarria A (2013) Experimental and theoretical evaluation of semicarbazones and thiosemicarbazones as organic corrosion inhibitors. Corros Sci 67:281–291
46. Mahgoub FM, Al-Rashdi SM (2016) Investigate the corrosion inhibition of mild steel in sulfuric acid solution by thiosemicarbazide. Open J Phys Chem 6:54–66
47. Ameer MA, Khamis E, Al-Senani G (2000) Adsorption studies of the effect of thiosemicarbazides on the corrosion of steel in phosphoric acid. Adsorpt Sci Technol 18:177–194
48. El-Shafei AA, Moussa MNH, El-Far AA (2001) The corrosion inhibition character of thiosemicarbazide and its derivatives for C-steel in hydrochloric acid solution. Mater Chem Phys 70:175–180
49. Ramya K, Mohan R, Anupama KK, Joseph A (2015) Electrochemical and theoretical studies on the synergistic interaction and corrosion inhibition of alkyl benzimidazoles and thiosemicarbazide pair on mild steel in hydrochloric acid. Mater Chem Phys 149–150:632–647
50. Musa* AY, Kadhum AAH, Mohamad AB, Takriff MS (2011) Molecular dynamics and quantum chemical calculation studies on 4,4-dimethyl-3-thiosemicarbazide as corrosion inhibitor in 2.5 M H_2SO_4. Mater Chem Phys 129:660–665
51. Badr GE (2009) The role of some thiosemicarbazide derivatives as corrosion inhibitors for C-steel in acidic media. Corros Sci 51:2529–2536
52. Ebenso EE, Isabirye DA, Eddy NO (2010) Adsorption and quantum chemical studies on the inhibition potentials of some thiosemicarbazides for the corrosion of mild steel in acidic medium. Int J Mol Sci 11:2473–2498
53. Mohan P, Kalaignan GP (2013) 1, 4-Bis (2-nitrobenzylidene) thiosemicarbazide as effective corrosion inhibitor for mild steel. J Mater Sci Technol 29(11):1096–1100
54. Shahabi S, Norouzi P, Ganjali MR (2015) Electrochemical and theoretical study of the inhibition effect of two synthesized thiosemicarbazide derivatives on carbon steel corrosion in hydrochloric acid solution. RSC Adv 5:20838–20847
55. Wazzan NA (2015) DFT calculations of thiosemicarbazide, arylisothiocynates, and 1-aryl-2,5-dithiohydrazodicarbonamides as corrosion inhibitors of copper in an aqueous chloride solution. J Indus Eng Chem 26:291–308
56. Ita BI, Offiong OE (1999) Corrosion inhibitory properties of 4-phenylsemicarbazide and semicarbazide on mild steel in hydrochloric acid. Mater Chem Phys 59:179–184
57. Fouda AS, Madkour LH, Elshafei AA, Elasklany AH (1995) Inhibitory effect of some carbazides on corrosion of aluminium in hydrochloric acid and sodium hydroxide solutions. Mat-wiss u Werkstofftech 26:342–346
58. Ramya K, Anupama KK, Shainy KM, Joseph A (2017) Corrosion protection of mild steel in hydrochloric acid solution through the synergistic of alkylbenzimidazoles and semicarbazide pair—electroanalytical and computational studies. Egypt J Pet 26:421–437
59. Al-Bonayan AM (2015) Inhibiting effect of thiosemicarbazide and 4-phenyl thiosemicarbazide towards the corrosion of carbon steel in H_3PO_4 solutions. Int J Electrochem Sci 10:589–601

Ammonia from Steelworks

Agustin Valera-Medina and Alberto Roldan

Abstract Ammonia has been produced over the last centuries in several ways, with the Haber–Bosch process leading current production due to its efficiency and feasible deployment. However, previous to the leading positioning of the Haber–Bosch process, ammonia used to be manufactured using coal-based gas works. Coke, a remnant of the process, has been widely used for steel production processes, thus making reasonable the integration of these gas facilities into the production of steel for better economic profiles. Although this ammonia production process is currently used only in a minor share of the total ammonia market, there are locations where it is still employed to obtain the chemical for fertilizing applications. This chapter is dedicated to the production of ammonia from such steelworks, detailing some of the history, fundamental and current trends behind the process that set the foundations of ammonia as one of the main global chemicals. Steel, which will still be produced over decades, can indirectly provide a chemical that supports a more sustainable agenda if better process integration is achieved, minimizing emissions and energy losses.

Keywords Ammonia · Cove oven Gas · Steel · Mond · Energy

1 Background

Since the 1860s, ammonia obtained from the destructive distillation of coal was used as a source of nitrogen for fertilizer purposes. Ammonia liquor was recovered by sulfuric gas absorption that was employed after scrubbing coal gas with water. Ammonium sulfate, the resultant from the process, was then used as fertilizer [1]. It is estimated that approximately 140,000 tons/year of this chemical were produced around Europe by the end of the nineteenth century.

It was in 1889 that Ludwig Mond, a German-British chemist, discovered, during his search of a process to manufacture ammonium sulfate, that coal combustion

A. Valera-Medina (✉) · A. Roldan
College of Physical Sciences and Engineering, Cardiff University, Wales, UK
e-mail: valeramedinaa1@cardiff.ac.uk

Inamuddin et al. (eds.), *Sustainable Ammonia Production*,
Green Energy and Technology, https://doi.org/10.1007/978-3-030-35106-9_4

could produce ammonia when the reaction takes place with air and steam. The produced gas, in combination with other species, was named Mond gas [2]. His findings suggested that low-quality coal could react with superheated steam producing high valuable gas. Dilute sulfuric acid sprays were then employed to remove the ammonia, forming ammonium sulfate in the process. Different from other processes, Mond's modification was based on the restriction of the supplied air, which was then filled with steam to generate low-temperature atmospheres. Since these low temperatures were lower than the temperatures required to dissociate ammonia, the recovery of the latter was maximized [2]. Mond realized at the time the great potential of this process to produce ammonia to satisfy the fertilizer market, although his initial interest was to ensure the supply of ammonia to his alkali factory.

Mond's process was utilized at Brunner from 1902, which was followed by implementation at various other sites in Britain, Argentina, Spain and the USA by 1903. These plants required massive capital investment to be economically viable. Over 182 tonnes of coal per week were employed to produce ammonia profitably with efficiencies as high as 80% [3]. It is estimated that the production of British ammonium sulfate reached ~220,000 tons/year by 1902, with 68% obtained from gas works and 9% from coke, carbonizing and Mond gas works [1]. Two-thirds of the total production were exported across the globe and used as fertilizer. An important destination for British fertilizer was Japan until the beginning of World War I [1]. By then, the production of ammonium sulfate was estimated at around 270,000 tons/year [4].

Further technological advances took place across Europe driven by companies such as Semet and Evence Coppee that developed new techniques to increase the ammonia recovery from coke ovens. Amongst other leaders that worked on this subject, Heinrich Koppers introduced the coke oven process into the US industry in 1907. Establishing Heinrich Koppers AG in 1904, he created the so-called half direct process for recovery of tars and ammonia from coke oven gas. By 1923, 90% of all coke oven plants in the Ruhr district used this process [5].

Before World War II, significant quantities of ammonia were manufactured from derived coke producing processes via coal conversion. Coal-based ammonia production dominated the industry until then [6]. It is estimated that around 2.7 million tons of ammonia per year were produced in this manner [7]. However, the situation would change by 1960, where most of the ammonia (by this time, 16 million tons per year) was obtained using natural gas from which hydrogen was obtained. Close to the end of the last century, coal and coke only accounted for 10.9% of the feedstock to produce ammonia. Nowadays, some industries still produce ammonia from steelwork gases using these processes, although most have closed down from their glorius years, Fig. 1. Examples are the fertilizer plants located in India at Durgapur (West Bengal), Rourkela (Orissa) and Bhilai (Madhya Pradesh) and those at Burnpur and Kulti (West Bengal) [8].

However, with the technological advancements and for the first time, some companies have returned to the use of steelwork gases for the production of ammonia. ThyssenKrupp recently announced the production of ammonia using steel mill gases [9]. Findings obtained through the Carbon2Chem project in collaboration with Fraunhofer-Gesellschaft, the Max Planck Society and 15 other partners (with an

Fig. 1 Former South Staffordshire Mond gas company works [3]. Courtesy of the National Grid Archive, UK

estimated €60M budget) revealed that ThyssenKrupp succeeded in producing pure chemicals from these steel waste streams in 2018. The first ammonia production took place in the Carbon2Chem technical centre in Duisburg, a pilot plant in which laboratory results were validated under practical industrial conditions using gases from regular steel mill operations [9]. The steel mill gas is comprised of 44% nitrogen, 23% carbon monoxide, 21% carbon dioxide, 10% hydrogen and 2% methane (vol), thus making it suitable for the production of carbon and hydrogen synthetic gases such as ammonia. The solution developed in Duisburg could be transferred to over 50 steel mills worldwide [9]. This scenario, combined with new energy storage trends, opens again the possibility of producing ammonia from steelworks through reduced cost and increased efficiencies, whilst minimizing emissions and reducing energy losses for more sustainable and integrated processes.

2 Production

Coke oven gas (COG) is a mixture of gases produced in steel plants during the carbonization of coal. It contains a great variety of gases, which vary depending on the process and coal source. However, concentrations tend to be around 50% H_2, 25% CH_4, 8% CO, 6% N_2 and minor amounts of hydrocarbons, CO_2 and impurities (i.e. ammonia, hydrogen sulfide, hydrogen cyanide, ammonium chloride, benzene, toluene, xylene, naphthalene, tars and carbon disulfide, in volume) [6, 10]. Coke oven gas is still employed to produce ammonia by steam reforming or by separation of hydrogen by cryogenic processes [10]. Before these processes, the precipitation of tar is required to avoid corrosion and unwanted emissions [11]. The cryogenic processes involve liquid nitrogen wash [6, 8], which is obtained from air separation units to produce oxygen for the steelworks [7]. The combination of nitrogen and hydrogen, as in the following reaction, yields ammonia [12].

$$N_2 + H_2 \rightarrow NH_3 \ \Delta_r H^\circ = -45.9 \text{ kJ/mol}$$

The general principle observed in the sequence of operation in the by-product plant is to follow the tar/liquor separation from the gas by sequential removal of components [4]. The operation is usually carried out at positive pressures (70–150 mBar) to avoid air in-leaks. The normal sequence of operation is as follows [10],

1. Tar and liquor separation
2. Primary gas cooling
3. Compression in exhausters
4. Electrostatic tar droplet removal
5. Secondary/final gas cooling
6. Ammonia removal
7. Benzole removal
8. Naphthalene removal
9. Hydrogen sulfide removal.

Sequences may vary depending on the by-product plant and any specialized processes.

3 Ammonia Removal Processes

The retrieval of by-products from steelworks is significant as there are many potential benefits. Nevertheless, COG needs to be treated and purified of contaminants that have the potential to foul and corrode pipeworks [10, 13, 14]. Yields from this treatment depend on several factors, mainly related to the carbonization conditions [10].

Historically, COG by-product constituents were valuable, often more so than the coke itself, particularly within the agricultural sector (i.e. fertilizers). Currently, the emphasis on the by-product plant to purify COG became more about the treatment of the gas to produce environmentally clean fuel and less about the recovery of by-products. There has been a significant focus on optimizing the industrial purification process of coke oven gas in order to achieve cost-efficient and environmentally friendly practices. Germany, in particular, has developed coke plants that have the highest environmental standards in comparison with other coking facilities. Nevertheless, there is still further research being carried out to improve the efficiency and environmental standards of coke plants across the globe [14]. Amongst the COG components, the recovery of ammonia is of particular interest due to economic and environmental reasons.

As mentioned above, COG is a complex gas with many constituents, some of which are useful for many applications; for example, hydrogen, methane and carbon monoxide are retained for the production of steel and production of extra power; paraffinic and unsaturated gases are also retained for further uses [10, 14]. Small traces of harmless, inert gases such as nitrogen or CO_2 are also kept in the final gas [10]. Ammonia removal from both COG and the flushing liquor is now a universal requirement of coke ovens as it is corrosive to carbon steel pipework, forming nitrogen oxides during oxidation, which are detrimental to the environment due to the over-nitrification of groundwaters and soils besides the formation of NO_x noxious gases. However, this removed ammonia can potentially be used for other purposes, for example, power or combined fertilizer/energy storage concepts. Methods presented in this section could be used in a more integrated manner to recover ammonia for energy storage, power and heating, raising overall efficiency whilst reducing losses.

3.1 Ammonia Removal from Flushing Liquors

Conventionally, excess flushing liquor was commonly used to quench the hot coke. However, this is no longer an acceptable practice for environmental reasons. Alternatively, the flushing liquor distillation followed by further treatment in a biological effluent treatment (BET) plant would remove the ammonia content. The distillation stage reduces operational costs, for example, by feeding the flushing liquor to a distillation column and feeding a counter-current flow of stripping stream beneath it, which causes the ammonia to vaporize out of the overhead vapours before further treatment. The treated liquor is then cooled and passed onto the BET plant for further ammonia removal by decomposing, for instance, ammonium chloride and ammonium sulfate, upon alkali addition.

Traditionally, calcium hydroxide was used as the alkali source as it was inexpensive and readily available. However, the formation of insoluble calcium salts caused fouling problems. Alternatively, sodium carbonate is used as insoluble salts were not formed, although the impurities present caused similar problems of fouling. Currently, sodium hydroxide is used, and although it is more expensive than traditional alkaline salts, it outweighs previous fouling limitations.

3.2 Ammonia Removal from COG

An estimated 2.5–3.0 kg of ammonia is produced per ton of coal employed for the production of coke due to the high temperatures of the process. The current technology to remove ammonia from COG streams is well established, Fig. 2, and

Fig. 2 Tata Steel Port Talbot, UK. Although not commercialized anymore, ammonia traces from COG processing can potentially be used for new purposes. Courtesy of Tata Steel, UK

although the so-called water-wash is the more conventionally method, other methods such as the ammonium sulfate and Phosam processes are also employed [10, 13].

The **ammonium sulfate method** involves the retention of NH_3 in solution by the addition of sulfuric acid (H_2SO_4) generating ammonium sulfate ($(NH_4)_2SO_4$) [10].

$$NH_3 + H_2SO_4 \rightarrow (NH_4)HSO_4 + NH_3 \rightarrow (NH_4)_2SO_4$$

The $NH_4(HSO_4)$ is recovered by crystallization before the formed crystals are centrifuged, washed, and then dried. This method of NH_3 recovery differs in practice depending on the type of gas/liquor contacting device and the equipment used for crystallization. There is, however, an economic disadvantage with the ammonium sulfate process; the price of H_2SO_4 required to form $(NH_4)_2SO_4$ can cost twice as much as the value of the produced ammonium sulfate crystals.

As previously stated, ammonia or ammonium sulfate scrubbed from COG gas is still employed as fertilizer in various places across the world. There are three methods used for this recovery: direct, indirect and semi-direct [6, 7]. These are described in Table 1.

The **Phosam method** for ammonia recovery was developed in the USA to produce pure anhydrous NH_3. The high-value product means that this process is much more economically favourable than the ammonium sulfate processes. The Phosam process selectively absorbs NH_3 from COG via direct contact with an aqueous solution of phosphate, which is added only in minimal quantities to the reaction. This process has very high efficiency achieving >99% recovery of NH_3 from the COG [16]. Through the reaction,

$$NH_3 + H_3PO_4 \rightarrow (NH_4)H_2PO_4 + NH_3 \rightarrow (NH_4)_2HPO_4 + NH_3 \rightarrow (NH_4)_3PO_4$$

Table 1 Ammonium sulfate scrubbing methods [6, 7, 15]

Method	Description	Advantages	Disadvantages
Direct	Tar is removed by cooling down the inlet gas. The gas is then passed through a saturator and washed with H_2SO_4. Ammonium sulfate is produced, centrifuged, washed and dried	High recovery of effluents with low investment and reduced operating expenses	Products could be contaminated (i.e. tar, pyridines and chloride). If the reactor is used as scrubber and crystallizer, the pH is difficult to handle to reduce impurities. Chloride from fuel or water could also react with the mixture and generate corrosion problems
Indirect	The gas is cooled and liquefied. Liquors are passed through a bubble-cap still. "Free" ammonia from the salts is released when the liquors get in contact with steam, whilst posterior treatment through lime decomposes the "fixed" ammonium salts. Ammonia can then be stripped with water or combined with sulfuric acid to form ammonium sulfate	The method has considerable flexibility, with a product (i.e. ammonium sulfate) free of impurities	The effluent disposal is highly problematic, with high ammonia losses because of the reduced reaction and absorption. Due to the higher complexity, the operation is more expensive than with the direct method
Semi-direct	This method offers a solution that incorporates concepts from both the indirect and direct processes. The gas is cooled, and traces of tar are removed. Aqueous condensates are sent to ammonia spill. The mainstream and the released ammonia are combined and heated up to 70 °C. The gas is scrubbed with a nearly saturated ammonium sulfate solution comprised of 5–6% H_2SO_4. This last part of the process takes place at temperatures between 50 and 70 °C	The method is the preferred option for extensive facilities, with more significant ammonia recovery and salts free from impurities	Although superior to the direct and indirect methods, the water balance during the saturation operation needs to be carefully controlled to enable the proper reaction of species. Moreover, salt incrustation could lead to maintenance problems

The **water-wash method** involves the use of water to strip contaminants from COG [13]. Aqueous absorption liquor is fed into an NH_3 washer vessel in a counter-current flow to the COG, leading to an ammonia solution of high concentration. The following step is distillation and recirculation of the condensed phase, which usually is biologically treated to eliminate any residual NH_3.

The washer vessel is usually placed after the tar precipitator in the by-product plant. The COG requires cooling before it enters the washer. Since this part of the process is temperature dependent, it has been found that NH_3 removal is more efficient at low temperatures. Indeed, the solution of gases such as NH_3 depends on the liquid phase temperature according to Henry's law of solubility. However, in the steelwork industry, the washer must not be operated at lower temperatures than the outlet temperature of the gas cooling stage in order to avoid fouling by naphthalene precipitation.

Alternative new technologies such as membranes are easing even further the removal of ammonia from flushing liquors [17]. Membrane distillation is being investigated worldwide as a highly efficient and affordable technology [17, 18]. Hydrophobic membranes (flat-sheet, hollow fibre and spiral wound) are preferred for ammonia extraction due to their hydrophobic characteristics, excellent organic resistance and chemical stability with acidic and alkaline solutions [17]. The strong hydrophobicity of the membrane prevents liquid transportation through it, thus facilitating the separation of species with different vapour pressures [19]. The partial pressure gradients across the membrane result in the transfer of the volatiles from the liquid phase to the vapour phase. As the temperature gradient is maintained across the membrane, the transport of water vapour occurs continuously. Meanwhile, other species remain on the other side of the membrane, thus separating water from the mixture. Conventional flat-sheet porous has been applied for membrane distillation with efficiencies varying between 70 and 90% [20].

4 Utilization of Ammonia from Steelworks

Ammonia from steelworks can be used in many ways. Similar to ammonia produced by any other method, the versatility of the chemical enables its use for fertilizing, heat production and chemical process applications. However, different from other processes, ammonia from steelworks can be employed using gaseous waste streams only available onsite to reduce cost or to generate extra heat/power for processes needed to produce steel. Therefore, this section is dedicated to evaluating some of these specific processes.

4.1 Ammonium Sulfate

Ammonium sulfate can be used as a fertilizer. Although its use for this application is relatively smaller when compared to urea, ammonium nitrate solutions and anhydrous ammonia [6], ammonium sulfate can still be employed as a fertilizer if available. The manufacturing and delivery method will vary depending on the customer's specifications.

4.2 Incineration

Incineration of ammonia vapours product of the coke oven gas process has been a common practice since the 1960s [4]. Ammonia, with a low heating value (LHV) of 18 MJ/kg, can be burned using a pilot flame without the need of other doping agents [21]. Companies around the globe employ these systems to reduce costs and mitigate contaminants, many of which are intrinsically dedicated to the production of steel [22]. However, the process carries out the production of emissions such as NO_x and SO_x, which are extremely detrimental to the atmosphere. Since ammonia contains nitrogen, its combustion at elevated temperatures generates NO_x emissions higher than those established for environmental regulatory purposes, requiring the implementation of novel combustion techniques to reduce temperatures whilst burning chemicals such as sulfur content species.

Therefore, the presence of these unwanted emissions in the stack has led to a close review of the process, with new installations now being used only for standby or emergency applications as short-term measures, i.e. for maintenance or unstable operation [4]. Moreover, some companies have looked at incineration as a possible solution to some of their heating requirements. Specialized steel production, which employs ammonia as part of their process, has attempted to utilize ammonia-containing gases for additional heating applications [23].

4.3 Concentrated Ammonia Liquor

Ammonia liquor can be produced from various ammonia vapour streams or water-washed methods. A preliminary condensation is carried out to minimize water content. The condensate is returned to the distillation column, with the remaining vapours then condensed to produce ammonia liquor [4]. Ammonia liquor produced by this method has a limited concentration of 15–20% ammonia, a consequence of the ammonium carbonate that is also formed on the condenser surface. This problem can be reduced by reducing acid gases such as carbon dioxide by "deacidification" processes via steam stripping [4].

A standard method to cool and condense ammonia vapours is by direct contact with recirculated cooled ammonia, thus avoiding localized sub-cooling that can lead to crystallization and fouling. The production of concentrated ammonia liquor is also used to provide a low-cost standby in plants where the regular ammonia handling facilities are out of service for maintenance [4].

Another option is to use the concentrated ammonia liquor or anhydrous ammonia, i.e. obtained from the former, for power purposes. Hewlett et al. [13] examined the potential of using ammonia in both its ammonia vapour (AV) and recovered anhydrous ammonia (AA) forms, to produce power using gas turbine (GT) technology. The priority of the study was to minimize NO_x concentrations and other harmful emissions whilst optimizing operational performance. After analyses using 0-D, 1-D and ASPEN software with various combinations of ammonia and COG blends, the group concluded that ammonia from steelworks could be used to produce power >4 MW at efficiencies up to 46%. Further information can be found in [13]. Figure 3 shows a representative flame obtained with such blends. According to numerical analyses, adding 15% COG to both the AV and AA gives the optimal balance of reactivity and lower pollutant products. Therefore, the potential of using ammonia product from coke processes might have a viable application to support energy storage in steelwork complexes.

Fig. 3 Ammonia-COG flame

4.4 Catalytic Destruction

Another option presented by the industry was to destroy ammonia by cracking the molecule. Instead of producing contaminants (NO_x, SO_x) or adding extra costs to the production of ammonium sulfate, the concept seeks to recover hydrogen from the ammonia molecule, therefore enabling the use of hydrogen in other parts of the process as a doping agent to increase combustion performance.

The process employs COG that, in conjunction with vapours, air and ammonia are fed to a burner with high flowrate controls. The reaction favours the combustion of COG, thus creating an atmosphere consisting of ammonia vapours and COG combustion products at ~1150 °C. A catalytic nickel-based bed is used, which cracks the ammonia into hydrogen and nitrogen. Hydrogen cyanide reacts with the water vapour producing nitrogen, hydrogen and CO_2. The gases then pass through a waste heat boiler to be cooled down to 300 °C. The gases may pass directly to the raw COG main, thus closing the loop, or they may be further cooled by water for other purposes [4]. This process enables a reduction of NO_x or SO_x emissions.

4.5 Final Remarks

Ammonia obtained from steelworks used to be the primary source for the production of fertilizers before the end of the first half of the twentieth century. However, due to the increasing introduction of natural gas and its cleaner processing, ammonia from steelworks was relegated, now sharing less than 12% of the overall ammonia production worldwide. Although some developing economies still use the concept for fertilizing applications, costs and environmental regulations have led to a decline of ammonia production via this method.

However, since steel is a significant component of today's society, the use of ammonia from the generation of coke oven gas or steel mills can still be a potential source for chemical storage that could be implemented for backup applications in power and/or heating auxiliary processes. Processes that are used to remove NH_3 could be integrated with other developments to raise efficiency and mitigate unwanted emissions. Current studies and various industrial projects show that ammonia, recovered from waste streams found in the steel production chain, can have significant advantages for energy support and chemical production. Thus, further progress on the topic should be expected in the years to come, as the use of these waste streams in an integrated manner can raise efficiencies whilst mitigating emissions, thus contributing to the sustainability steelworks.

Acknowledgements Cardiff University gratefully acknowledges the support from the Welsh European Funding Office (WEFO) through its programme "Flexible Integrated Energy Systems (FLEXIS)", project no. 80835. REFERENCES NEED TO BE AMENDED. - REF 3 HAS "COMMAS", AND INSTEAD OF HAVING [ONLINE] IT SAYS -AVAILABLE ONLINE-, REF 5 ONLY 1 "COMMA".

References

1. Travis AS (2018) Nitrogen capture. The growth of an international industry (1900–1940). Springer, 410 p
2. R.D. Wood and Co. Philadelphia (1903) Mond gas. [Online] https://archive.org/stream/mondgas00woodrich#page/n5/mode/2up. Accessed 14 Feb 2019
3. CLAIRE (2014) "Gasworks profile D: produced gas plants" supported by National Gas Grid and Parsons Brinckerhoff. Available Online https://www.claire.co.uk/component/phocadownload/category/9-other-cl-aire-documents?download=433:gasworks-profile-d-producer-gas-plants. Accessed 9 May 2019
4. Vroomen H (2013) The history of ammonia to 2012. [Online]. By The Fertilizer Institute. Available in www.firt.org/sites/default/files/2Vroomen.pdf. Accessed 20 March 2019
5. Dierschke A (1955) "Development of the coke oven industry"[Online]. By the JSTAGE. Available in https://www.jstage.jst.go.jp/article/jie1922/35/3/35_3_157/_pdf. Accessed 20 Mar 2019
6. Vasant G, Krishnamurthy VN, Gowariker S, Dhanorkar M, Paranjape K (2009) The fertilizer encyclopedia. Wiley, 860 p
7. United Nations Industrial Development Organization (UNIDO) (1998) Fertilizer manual, 2nd edn. Kluwer Academic Publishers, 619 p
8. Kolay AK (2007) Manures and fertilizers. Atlantic Publishers and Distributors, Weinheim, Germany, p 55
9. BioAge Group. ThyssenKrupp produces ammonia from steel mill gases; Carbon2Chem. Available in Green Car Congress [Online] www.greencarcongress.com. Accessed 14 Feb 2019
10. Wright K (2005) Coke oven gas treatment. Tar, liquor, ammonia. In: The coke oven manager's year book, pp 221–257
11. Caillat S (2017) Burners in the steel industry: utilization of by-product combustion gases in reheating furnaces and annealing lines. Energy Procedia 120:20–27
12. Uribe-Soto W, Portha J-F, Commenge J-M, Falk L (2017) A review of thermochemical processes and technologies to use steelworks off-gases. Renew Sustain Energy Rev 74:809–823
13. Hewlett SG, Valera-Medina A, Pugh DG, Bowen PJ. Gas turbine co-firing of steelworks ammonia with coke oven gas or methane: A fundamental and cycle analysis. In: Proceedings of the ASME turbo expo 2019, Phoenix, USA. Ref. No. GT2019-91404
14. Razzaq R, Li C, Zhang S (2013) Catalytic methanation of CO and CO_2 in coke oven gas over Ni-Co/ZrO_2-CeO_2. Fuel 113:287–299
15. Sweeney MP (1953) Coke oven by-product recovery process. Patent US2943911A. Available Online https://patents.google.com/patent/US2943911. Accessed 23 May 2019
16. Ratnayaka DD, Brandt MJ, Johnson KM (2009) Water supply, 6th edn. Butterworth-Heinemann, 744 p
17. Kunz A, Mukhtatar S (2016) Hydrophobic membrane technology for ammonia extraction from wastewaters. Engenharia Agricola 36:377–386
18. Zhao ZP, Xu L, Shang X, Chen K (2013) Water regeneration from human urine by vacuum membrane distillation and analysis of membrane fouling characteristics. Sep Purif Technol 118:369–376
19. Udert KM, Wächter M (2012) Complete nutrient recovery from source-separated urine by nitrification and distillation. Water Res 46:453–464
20. El Bourawi MS, Khayet M, Ma R, Ding Z, Li Z, Zhang X (2007) Application of vacuum membrane distillation for ammonia removal. J Membr Sci 301(1–2):200–209
21. Valera-Medina A, Xiao H, Owen-Jones M, David WIF, Bowen PJ (2018) Ammonia for power. Prog Energy Combust Sci 69:63–102
22. Masons P. The port talbot steelworks (Power Generation Enhancement) order. TATA Steel, Report MAH1.01, p 211
23. Fatla OHM, Valera-Medina A, Robinson F, Cichuta M, Beynon N (2018) Development of convection in high temperature coil annealing furnaces using rotating cylinder technique. Appl Therm Eng 129:1392–1402

Catalytic Ammonia Decomposition for Hydrogen Production: Utilization of Ammonia in a Fuel Cell

Lateef Jolaoso and Sharif Fakhruz Zaman

Abstract Ammonia is one of the best potential hydrogen storage materials, having a high volumetric (121 kg H_2/m^3) and gravimetric (17.75 wt%) hydrogen capacity. Its properties fully correspond to the DOE's (Department of Environment, USA) hydrogen storage requirements as a commercial hydrogen storage material. Ammonia can be used for onboard clean (CO_X free) hydrogen generation ($2NH_3 \Leftrightarrow N_2 + 3H_2$) for fuel cell-driven vehicles. The main challenge of using ammonia to produce clean hydrogen via an onboard catalytic decomposition process necessitates a catalyst able to decompose 100% ammonia at a low temperature (≥400 °C) and supply pure hydrogen to the fuel cell. Currently, only ruthenium-based catalysts showed activity to complete decomposition of ammonia at 400 °C and above but the scarcity of precious ruthenium put an economic constraint in the wide application of ruthenium-based catalysts and drive researchers to look for alternative (non-precious) catalytic materials for this reaction. This chapter describes briefly about hydrogen and current hydrogen production and storage technologies, the cost of hydrogen production from different processes, ammonia and current status of ammonia production followed by a detailed discussion of different ammonia decomposition catalysts.

Keywords Hydrogen · Ammonia decomposition · Transition metal catalysts · Bimetallic catalysts

1 Introduction

Our world is currently facing the greatest challenge of adverse climate change mainly for anthropogenic causes, i.e., excessive and inefficient use of fossil fuel (coal, oil, and natural gas) and building up of CO_2 in the atmosphere instigating global warming [1, 2]. For the adoption of an International Climate Agreement, there have been frequent meetings among countries since 2011 under the UN framework. In 2015 at the 21st conference of the parties in Paris, an agreement was made to make a shift in

L. Jolaoso · S. F. Zaman (✉)
Department of Chemical and Materials Engineering, Faculty of Engineering, King Abdulaziz University, Jeddah, Saudi Arabia
e-mail: sfzaman@kau.edu.sa

Inamuddin et al. (eds.), *Sustainable Ammonia Production*,
Green Energy and Technology, https://doi.org/10.1007/978-3-030-35106-9_5

social and economic activities toward sustainable and climate-friendly actions [3]. In the sequel to this, researchers are working to develop a CO_x-free alternative fuel and energy vectors that would serve the current and future global energy demands for transportation vehicles as around 26% CO_2 emission is from the transportation sector. There is a need for the replacement of fossil fuels as an energy source in transportation vehicles. Hydrogen (H_2) is one of the most potential alternatives and environmentally friendly transportation fuel, termed as the next-generation energy carrier when coupled with a fuel cell, in which the efficiency of the engine can be raised to more than 60% while the conventional internal combustion engines run on less than 30% efficiency. The evolution and adoption of the fuel cell as a replacement for the battery have grown considerably in the twenty-first century. This is due to its ability to continuously supply electricity from an external hydrogen fuel source in comparison with the limited energy storage capacity of a battery which would also require a long time to recharge. Despite the preponderant availability of the battery technologies and its use, manufacturing it is still expensive.

The problem of storage and transportation of hydrogen is yet another significant challenge for its feasible utilization as fuel [4]. The present high pressure—$\approx$700 bar—hydrogen storage is energy demanding and extremely expensive which causes restrictions on tank dimensions [5], and this is a big challenge in the success of hydrogen-powered vehicles. In proffering solution to this setback, huge research is going on to search for a suitable hydrogen storage material. Ammonia has proven to be an excellent choice of hydrogen source as it does not generate any CO_x ($x = 1, 2$) and the unconverted ammonia (NH_3) can be lessened to below 200 ppb level using appropriate absorber [6] and can be cracked catalytically for in situ generation of hydrogen for a small/medium vehicle powered by a fuel cell. This chapter describes briefly about hydrogen and current hydrogen production and storage technologies, the cost of hydrogen production from different processes, ammonia and current status of ammonia production followed by a detailed discussion of different ammonia decomposition catalysts.

2 Hydrogen

Hydrogen is the simplest and most abundant of all the elements on earth and is the lightest gas in the universe with a very high gravimetric energy density of 141 kJ/g compared to fossil-derived fuels. It does not exist alone but usually combines with other elements like oxygen as water and carbon as hydrocarbons [7]. It is a major constituent of natural gas (CH_4 is about 95% of the natural gas), oil, water and also present in all living organisms [8]. Biomass (cellulose) also contains hydrogen. There are different primary energy sources, such as nuclear, solar, wind, hydropower, and geothermal, with which the extraction of hydrogen can be done in a sustainable carbon freeway mainly from water splitting. Diverse sources of hydrogen (fossil, biomass, water, etc.) have made its production easy virtually in all parts of the world [9]. Figure 1 shows the energy sources from where hydrogen can be extracted. After

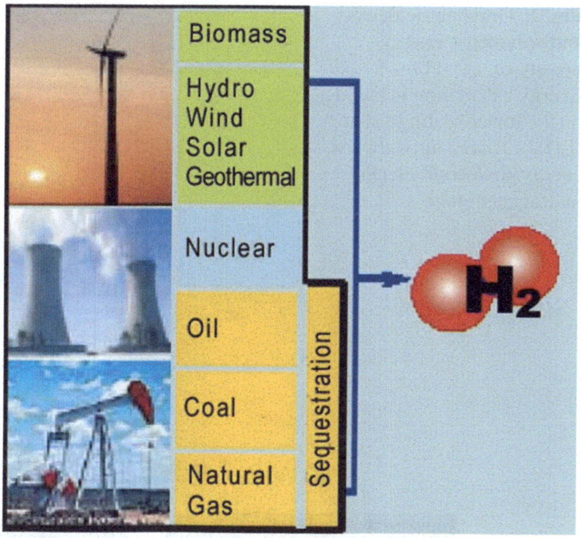

Fig. 1 Energy sources for hydrogen extraction. *Source* US Department of Energy

the production of molecular hydrogen from its feedstocks, the present energy within the molecule can be released, by reacting it with oxygen to form water. This can be realized either by the fuel cell or an internal combustion engine [8].

Production of hydrogen can be categorized based on its separation process from its feedstock. There are two major processes: the chemical process and the thermal process. The third one which is a biological process is under development.

Nearly all the produced hydrogen today (about 95% of produced hydrogen) uses the thermal process, i.e., steam reforming of natural gas and many other hydrocarbons. This method will continue to dominate even though there are some other methods yet to reach their peak of development. There are other processes such as

- Electrolytic processes
- Solar-driven processes
- Biological processes.

For a chemical fuel to be used in new technologies, its higher energy content should agree with its higher H/C ratio, and eventually, hydrogen is the most powerful of the chemical fuels [7] as shown in Fig. 2. These qualities resuscitated the idea of hydrogen economy in the 1990s and also in this era, fuel cell technology was growing to have efficient use of energy compared to the internal combustion engines [10]. The challenges of hydrogen to be used as a fuel in a fuel cell with its low volumetric energy density requires high pressure (300–700 bar) and or cryogenic system (−252.8 °C) to store hydrogen on board which also raise the safety issue. To meet these challenges, researchers are looking for a suitable hydrogen storage material, both chemical and physical, which has high hydrogen storage (gravimetric and volumetric) capacity and low cost of storage and generation. Different hydrogen storage methodologies are shown in Fig. 3. The featured target of hydrogen utilization

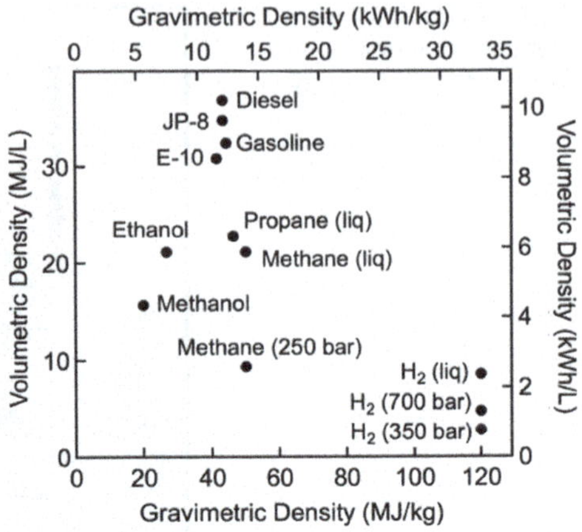

Fig. 2 Gravimetric density and volumetric energy density of several fuels. Energy calculation is based on the lower heating value (LHV). *Source* https://www.energy.gov/eere/fuelcells/hydrogen-storage

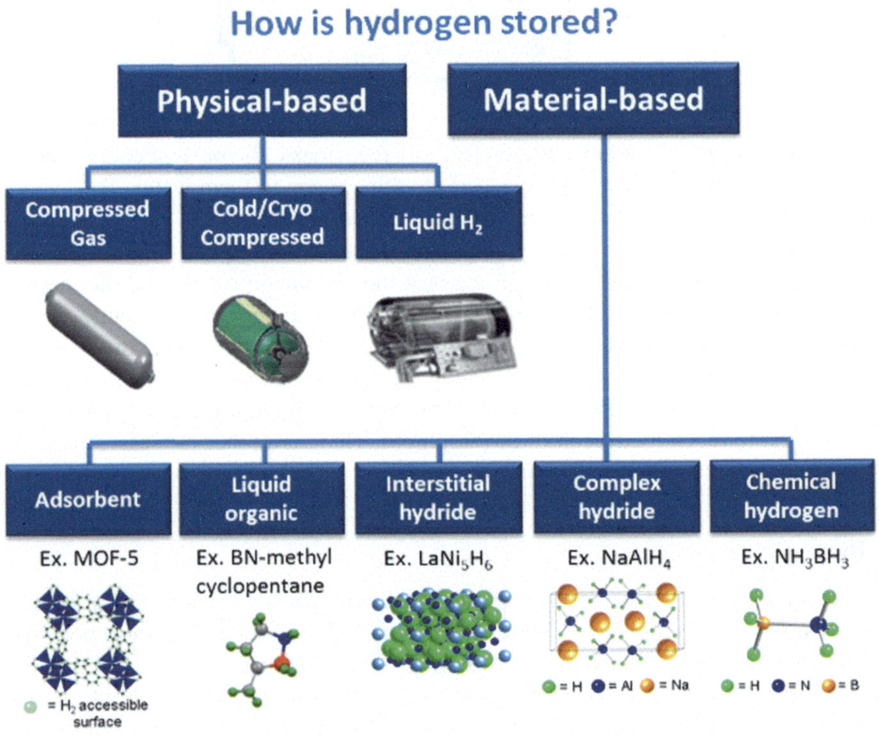

Fig. 3 Different hydrogen storage processes. *Source* https://www.energy.gov/eere/fuelcells/hydrogen-storage

is set by the US Department of Environment (DOE) in onboard light-duty vehicles, portable power generation, and material handling equipment.

3 Ammonia

Due to the inability of hydrogen to exist alone in nature and the challenges of storing and transporting it because of its small molecular size and low volumetric energy density, there is a need to find an alternative hydrogen vector [10]. Ammonia is one of the potential chemicals for storing and transporting hydrogen. It is a gas which is easy to liquefy and made up of three hydrogen molecules and one nitrogen molecule. It has a relatively high liquid density of 0.6 kg/l in comparison with that of hydrogen which is just 0.071 kg/l and the heat of combustion of ammonia is 11.2 MJ/l compared to 8.58 MJ/l for liquid H_2 [11].

In recent times, eminence attention has been gained by the concept of ammonia (NH_3) economy [12]. Similar to hydrogen, ammonia is not a greenhouse gas. Ammonia ranks as the second most extensively produced inorganic chemical commodity (Haber-Bosch process) in the world after sulfuric acid [13]. Over 130 million tonnes of ammonia is annually produced in the world, of which about 85% is used in fertilizer production [14]. The liquefaction of ammonia can be carried out at a pressure range of 8–10 bar at room temperature and can be stored following a similar storage technology as propane. The curve in Fig. 4 is a 2-Verhulst-functions fit for the global annual production of ammonia data. It is produced in over 80 countries,

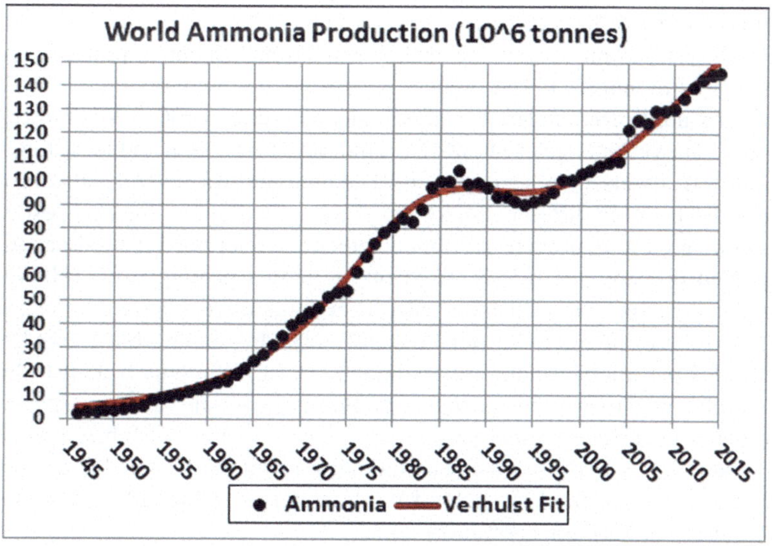

Fig. 4 2-Verhulst-functions fit for the global annual ammonia production [16]

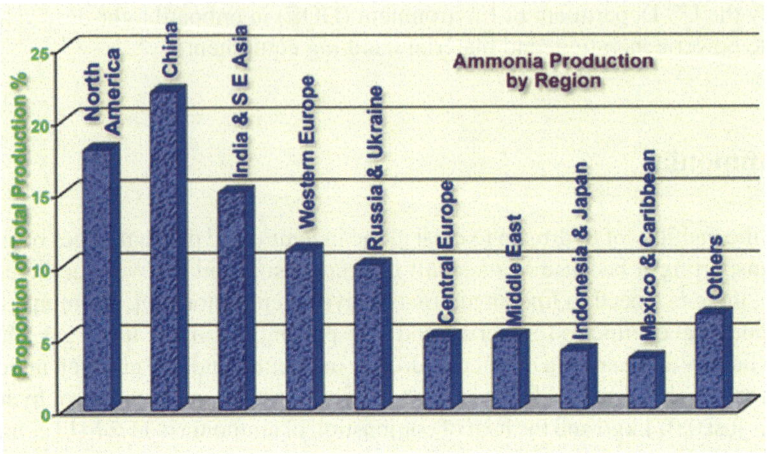

Fig. 5 Ammonia production by region [17]

and China is the highest producer of ammonia with 32.1% of the global production followed by India and Russia with 8.9% and 7.9%, respectively [15]. Figure 5 shows the production of ammonia by region.

The primary feedstock for ammonia is natural gas. However, there are lots of coal-based ammonia plants presently being used for ammonia production [18]. Ammonia, as a hydrogen storage material, would not only provide a short term need but also a long-term solution to the transportation sector with a projection of 200 years since there are already established and developed infrastructures with simultaneous sequestration of CO_2 for zero-carbon emission [19, 20].

The Haber-Bosch process is one of the most studied chemical processes for the synthesis of ammonia [21]. The process is performed at temperatures from 300 to 550 °C over a multi-promoted iron catalyst under pressures of 200–300 bar in a reactor having 2–4 catalyst beds gives about 15% ammonia conversion and recycling of the unconverted reactants (synthesis gas) [22]. Ammonia has several benefits which have been discussed in different books and journals, it is one of the cheapest energy sources compared to hydrogen, gasoline, methanol compressed natural gas (CNG), and liquefied propane gas (LPG) based on the higher heating value (HHV). The estimated energy (per kilowatt-hour) cost from ammonia is US$ 1.2, which is low compared to energy production from methanol and hydrogen which will cost US$ 3.8 and US$ 25.4 per kilowatt-hour, respectively [23]. So, ammonia is a cost-effective and viable chemical to be used as a hydrogen storage material. The cost of production of hydrogen from different sources and energy density is shown in Tables 1 and 2. The cost of production of hydrogen using ammonia cracking (with available technology) will be less if it is for the low scale of production (i.e., 10 m^3/h), which makes it viable for use it in light transportation vehicles.

Table 1 Hydrogen production cost from different sources [24]

Scale of H_2 production (m^3/h)	Cost of H_2 production, US$/($m^3/h$)			
	Water electrolysis	Natural gas reformation	Methanol reformation	Ammonia cracking
10	0.943	0.390	0.380	0.343
100	0.814	0.261	0.285	0.279
1000	0.739	0.186	0.226	0.241

Table 2 Hydrogen production cost from different sources with energy density [24]

Fuel	Price (USD)	Energy density (Wh/l)
Ammonia	425 per ton	4325
Hydrogen	10 per kg	2600
Natural gas	3.02 per MMBtu	6400
CPG	575 per MT	2500
LNG	0.16 per MT	7216
Methanol	1.5 per gallon	4600
Gasoline	2.49 per gallon	9700

4 Ammonia Decomposition for Hydrogen Production

The research on decomposition of ammonia is as old as its catalytic synthesis through the Haber-Bosch process in the early 1900s [25] which is the most studied catalytic reaction [26]. The traditional studies and works before the 1990s that were done on the decomposition of ammonia were purposefully to gain more insight about ammonia synthesis and its kinetics [27, 28]. But over the time, there was a gradual change since ammonia nowadays is well accepted as a promising vector of hydrogen and storage medium [12], i.e., using pure ammonia feed in the reactor to generate CO_x-free hydrogen [29]. Green, in 1982, was the first person to propose the production of carbon-free hydrogen through ammonia decomposition and to use it in a proton exchange membrane fuel cell (PEMFC) [30].

Some criteria were set by the US Department of Energy (DoE) for economically feasible hydrogen storage (physical storage system) in 2015. They are operational temperatures less than 60 °C, low-cost, speedy system filling, inert and non-toxic materials, high storage capacity with 81 g/L volumetric capacity and 9 wt% hydrogen content as shown in Fig. 6 [31]. Ammonia has proven to be an excellent choice of hydrogen source as it does not generate any CO_x ($x = 1, 2$) (CO is poisonous to the platinum electrode of the fuel cell) and the unconverted NH_3 can be reduced to below 200 ppb level using appropriate absorber [6, 29, 32]. It is easier to store compared to hydrogen being a liquid at 25 °C and pressure of 8 atm [33] and has well-established infrastructure (current gasoline tank in the transportation vehicles can be used to store ammonia) for its management and transportation [34]. In comparison with all carbonaceous compounds, ammonia possesses the highest gravimetric H_2

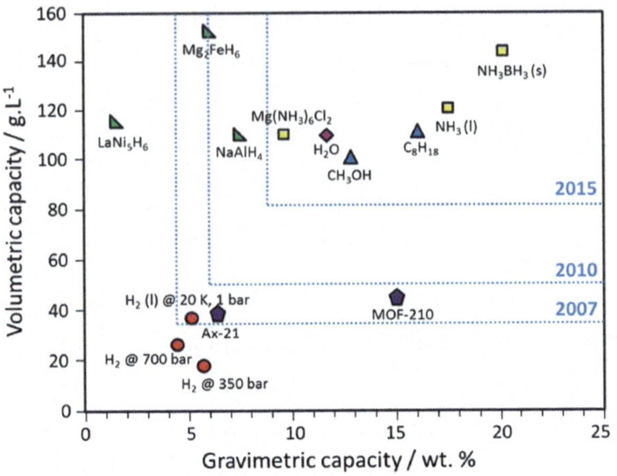

Fig. 6 Hydrogen storage capacity (volumetric and gravimetric) of different potential compounds. The dashed lines are the hydrogen storage criteria set by the US Department of Environment in different years [31]. Symbols: (1) circle = hydrogen under different conditions, (2) triangle = hydrocarbons, (3) pentagon = materials for H_2 physisorption, (4) right-angle triangle = metal hydrides, (5) diamond = water, (6) square = ammonia and related compounds

density, 17.75 wt% [35], and energy density of 3000 Wh/kg which is greater than those of methanol and other fuels [6]. Also, it has a very high volumetric H_2 density of 121 kgH_2/m^3 in the liquid form [5].

From the previous reports, it is well known that decomposition of ammonia can start to take place at temperatures over 300 °C [34]. And 99% of conversion is possible to achieve at atmospheric pressure at 400 °C attained from the thermodynamic study [35]. But catalytically much higher temperature is required. Different research efforts have been made to develop catalysts which are very active for this decomposition reaction. Different transition and noble metals, bimetallic alloys and mono-, bi-, and tri-metal carbides/nitrides have been investigated for ammonia decomposition as a catalyst [5, 6, 29, 31, 35] which will be discussed in the following sections.

5 Ammonia Decomposition Catalyst

5.1 Ruthenium-Based Catalysts

Ammonia decomposition study was mainly studied earlier to gain an insight into the ammonia synthesis reaction. Ruthenium-based catalysts are the most active at low temperate for this reaction and the most studied catalyst for this process. But the scarcity of this noble material put an economic constraint over the industrial use of the catalysts. Later, we have discussed other potential catalysts (low-cost transition metals and bimetallic and nitride- and carbide-based) for ammonia synthesis. The activity of the Ru-based catalysts is reported in Table 3. Figure 7 shows a volcano curve depicting the activities (N–H bond dissociation) of different transition metals used for ammonia decomposition with ruthenium at the apex of the curve.

Choudhary et al. [37] performed a comparison study of ruthenium, nickel, and iridium supported on silica for their catalytic activity for NH_3 cracking/decomposition and revealed that ruthenium has the highest rate of decomposition along with the highest rate of conversion of ammonia. They also compared the activity of ruthenium on different supports, i.e., silica and alumina, and found that the catalytic activity of silica-supported catalysts was higher than that of alumina-supported ones. Yin et al. [33] also studied support effect for the Ru-based catalysts and graded their catalytic activities as Ru/CNTs > Ru/MgO > Ru/TiO_2 > Ru/Al_2O_3 > Ru/ZrO > Ru/AC > Ru/ZrO-BD. They ascribed the excellent catalytic activity of Ru/CNTs to the high purity and graphitization of the CNTs and high dispersion of Ru particles over the support. Yin et al. also reported ammonia decomposition activity of ruthenium-based catalyst (Ru/CNTs) at 350 °C achieving a conversion of 6.31%, which was 8–40 times

Table 3 Activity of different Ru-based catalysts for NH_3 decomposition reaction

Catalyst	Temperature (°C)	GHSV (h^{-1})	Conversion (%)	Reference
Ru/K-CNT	400	150,000 ml	7.3	[6]
Ru/K-ZrO-BD	400	150,000 ml	5.3	[6]
Ru/CNT	400	60,000 ml	9.0	[36]
Ru/MgO-CNT	400	60,000 ml	13.0	[36]
Ru/ZrO_2	550	30,000 ml	77.0	[33]
Ru/SiO_2	600	30,000 ml	97.0	[37]

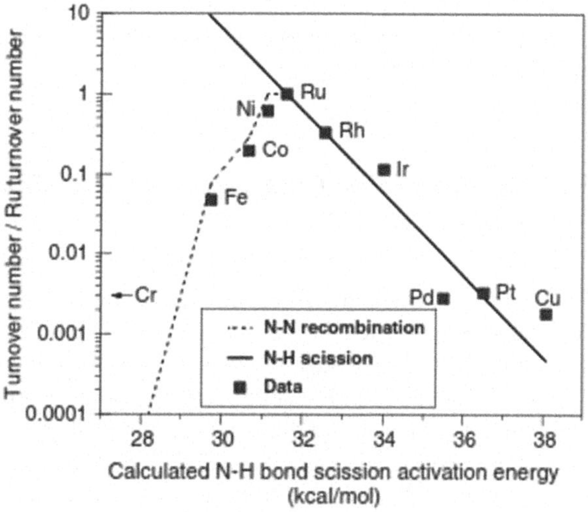

Fig. 7 A correlation between the rate of ammonia decomposition on several metals and the relative rate of N–H bond scission, and N–N recombination as estimated from the Blowers–Masel correlation [38]

greater than those studied over the supported Rh, Ni, Pt, Pd, and Fe catalysts at the same temperature with H_2 production rate of 2.11 mmol/(min g_{cat}).

Also, Li et al. [39] investigated the activity of Ru catalysts supported on different carbon supports for the decomposition reaction to gain an insight into the effect of carbon graphitization. They found out that graphite carbon (GC) to be the best support for ruthenium and ranked their catalytic activity in the following descending order; 5%Ru/GC > 5%Ru/CNTS > 5%Ru/CBS > 5%Ru/CB-C > 5%Ru/CMK-3 $\approx$ 5%Ru/AC. They concluded that the graphitic structure of the carbon-supported catalyst takes preference for better catalytic activity compared to the pore structures of the carbon supports.

Tsai and Weinberg [28] used the first principle method to study the steady-state decomposition of ammonia on Ru(001), they found that the rate-limiting step is temperature-dependent. At high temperatures, there is a competition between the ammonia desorption and N–H bond dissociation in chemisorbed ammonia while at low temperature, desorption of nitrogen controls the reaction. Egawa et al. [40] did further first principle investigation on this reaction on flat Ru(001) and stepped Ru(1110), and their results disclosed that over Ru(1110) nitrogen formation rate is an order of magnitude higher than that on Ru(001) in which the elimination of the edge effects have been done. Dietrich et al. [41] also reported a more open structure Ru(1121) and showed it to be more active than Ru(0001).

From experimental and theoretical investigations so far, Ruthenium is reported to be the most active catalyst component for ammonia decomposition which is enhanced when carbon materials, particularly carbon nanotubes (CNTs) are used as supports [6, 31, 42] and potassium hydroxide, KOH as a promoter [33]. At 737 K, Ru-K/CNT have 99.74% NH_3 conversion at a GHSV of 30,000 ml/(g_{cat} h). However, there are reports that claimed that cesium promoted ruthenium which is supported on graphite is the best and most active catalyst known [6]. Different findings also claimed that barium is a good promoter for ruthenium for ammonia decomposition [43], and recently, Wang et al. [44] claimed a superior performance of ruthenium supported on barium hexaaluminate having 99% conversion at 773 K at GHSV of 30,000 ml/(g_{cat} h).

5.2 *Monometallic Catalysts (Non-Ruthenium-Based Catalysts)*

From previous research, different monometallic (transition metal) systems of catalysts have been studied for the decomposition of ammonia for hydrogen production. The activity of these catalysts' is dependent on some other factors such as the conditions of decomposition reaction, and the catalyst supports and promoters. Thenard and Dulong [45] are the first to report metals that were recognized as catalysts for the decomposition of ammonia in 1823, and their activities are categorized in the following ascending: Pt < Au < Ag < Cu < Fe. Further, Fe, Ir, Rh, Pd, Ru, W, Ni,

Mo, Pt, carbides of molybdenum, vanadium, tungsten, and iron (MoC_x, VC_x, WC_x, FeC), nitrides of molybdenum, vanadium, and tungsten (MoN_x, VN_x, WN_x), as well as ZrON were studied for ammonia synthesis. Majority of the tested catalysts from the list showed good activity, but ruthenium and iron have been studied extensively in a plan to regenerate their activities toward ammonia synthesis [42]. For ammonia decomposition, the reactivity comparison for different transition metals are usually in the decreasing order as: Ru > Ir > Rh > Ni > Pt > Pd > Fe, these were reported by Choudhary et al. [37], Tanaka and Tamaru [46], Yin et al. [33] and Papapolymerou and Bontozoglou [47]. In a review by Bell and Torrent-Murciano [31], a more general activity trend involving additional metals was listed and having their activities in the following order Ru > Ni > Rh > Co > Ir > Fe > > Pt > Cr > Pd > Cu ≫ Te, Se, Pb, and this trend are based on using activated alumina as a support. Among them, cobalt, nickel, and iron were/are studied extensively for this process except for ruthenium and are discussed in the following sections.

5.3 Transition Metal (Co, Ni, Fe) Catalyst

Transition metals like Co, Ni and Fe are investigated as a potential catalyst for NH_3 decomposition [4] as an alternative to Ru as for their availability and low cost.

Cobalt-Based Catalysts

The investigation performed by Sorensen et al. [43] in which they compared four non-promoted transition metals, Co, Fe, Pd, and Ru supported on alumina, with one another, at temperatures ranged between 848 and 923 K and gas mixture having 50 vol.% of argon and ammonia each. It was established that Co has the best activity for NH_3 decomposition reaction among Co, Fe, and Ni. Zhang et al. [48] found that complete ammonia conversion was nearly reached over a CNTs-based catalyst containing 4.1 wt% Co at 973 K and flow of 2000 $cm^3\ g^{-1}\ h^{-1}$, thereby showing that commercial CNTs having Co nanoparticles to be very effective in the ammonia decomposition reaction. Lendzion-Bielun et al. [49] compared the industrial iron catalyst promoted with potassium and cobalt-based catalyst promoted with aluminum and calcium. They found that cobalt-containing catalysts have higher activity compared to iron-based one for ammonia decomposition, and this was apparent from, respectively, their activation energies of 138 and 111 kJ/mol for iron and cobalt catalysts. In a later work, Yin et al. [33] studied the effect of the hydrogen–ammonia gas compositional mixtures on the kinetics of the decomposition of ammonia over cobalt-based catalysts. They concluded that ammonia decomposition rate increases with an increase in the nitriding potential (*P*) which they defined as the ratio of ammonia partial pressure to hydrogen partial pressure with an optimum potential $P = 0.0019\ Pa^{-0.5}$ after which the decomposition rate was declined.

Different support materials have been used with cobalt, and it was found that cobalt catalyst is greatly influenced by them and their roles have been related to activity, stability, and cobalt particle size. Owing to the mechanical stability of carbon materials, they are the most studied supports for this reaction. They form an excellent metal-support with cobalt which brought about an improvement in the transfer of electron and thereby result in nitrogen desorption energy reduction [50]. Multi-walled carbon nanotube (MWCNT)-supported cobalt catalyst showed a better activity compared to that of corresponding nickel and iron [50]. The pre-treatment of these catalysts in nitrogen has been reported to improve its activity, and the pre-treatment temperatures also have a great impact on the cobalt particle size; i.e., the particle size increases as the temperature increases [51]. Zhao et al. [52] reported a case carbon CMK-3 treated with nitric acid; even though they did not test the activity of this material for the decomposition of ammonia, they found that the size of cobalt particles is also been controlled by the pre-treatment of the carbon materials with acid and this brought an improvement on the cobalt particles' dispersion on the carbon materials.

More works are still required to be done on cobalt-based catalysts for ammonia decomposition since no specific optimum cobalt particle size has been reported to give the best ammonia conversion [53, 54].

Iron-Based Catalysts

Iron is one of the early transition metals studied for the decomposition of ammonia to have an in-depth understanding of ammonia synthesis kinetics in the early time [55]. Significantly, iron promoted with alumina, silica, CaO, and K_2O at temperatures above 673 K is the most popularly used catalysts for large-scale ammonia synthesis. This has made iron catalysts to be widely investigated for the decomposition of ammonia [56]. Focused has been on ruthenium-based catalysts for ammonia decomposition but considering the cost–benefit, the bulk price of ruthenium is 50,000 higher than that of iron, thereby calls for the need to develop an iron-based catalyst with the high catalytic performance [39].

Very few investigations are reported on iron-based catalysts for ammonia decomposition mainly for its poor activity and stability in contrast to ruthenium-based catalysts. This was related to the very strong iron–nitrogen interaction unlike the ruthenium–nitrogen interaction [57], and this strong Fe–N interaction leads to sub-nitrides (stable nitrides) formation or even bulk nitrides, and this was considered in the early studies to be the inhibitors for the ammonia decomposition reaction and eventually deactivate the catalyst [31, 58]. To achieve high activity of the iron catalyst, the deactivation process could be reversed with high energy, whereby the large energy barrier for nitrogen desorption is overcome using high reaction temperature, which again brings about the sintering of iron particles and leads to irreversible deactivation of the catalyst due to the loss in active surface [59].

Several works have been performed to investigate the NH_3 decomposition kinetics over iron-based catalysts. In general, the rate-limiting step of this dissociation reaction is the combination of N atoms and desorption from the catalyst surface [31].

The decomposition of NH_3 over iron catalysts has led to the derivation of different types of rate equations with Temkin–Pyzhew equation being the most frequently used equation to describe the kinetics for ammonia decomposition [60, 61] as described below.

$$r = k\left(\frac{P_{NH_3}^2}{P_{H_2}^3}\right)^{0.25} \quad (1)$$

Arabczyk and Zamlynny [60] studied the rate of reaction of catalytic decomposition of NH_3 over iron nitride (Fe_4N) and clean iron (Fe) and compared their activities and the promoting effect of potassium on the catalysts. They found that reaction rate is directly proportional to ammonia concentration over iron nitride and fourfold faster compared to when the nitrides are covered with potassium; this was due to the strong Fe–N bond enthalpy that can bring about the formation of surface nitrides which is not suitable for ammonia decomposition. Also, the activation energy of clean iron is twofold higher than that of the iron nitride. Potassium on the catalyst surface decreased the activation barrier of the reaction.

Nickel-Based Catalysts

Another cost-effective transition metal for this decomposition reaction is nickel, which can be used as a catalyst and has attracted the attention of researchers as a substitute for ruthenium [31, 62]. Earlier studies on nickel as a catalyst mainly focused on ammonia decomposition kinetics as a supplementary study for understanding ammonia synthesis reaction mechanism. Majority of these reactions were done under some conditions for example under low ammonia concentration and very low pressure, or in a mixed condition, e.g., decomposition of ammonia with other reactions (in parallel) [62, 63].

It has been reported from the previous studies that Ni is a very active catalyst, e.g., Ni/Al_2O_3, for ammonia decomposition [29, 64] with the ammonia conversion reaching almost completion (greater than 99%) at atmospheric pressure and moderately high temperature (greater than 550 °C) [65].

The first N–H scission step energy, in NH_3 dissociation surface elementary reaction steps, for Ni-based catalysts is lower than that of iron- and cobalt-based catalysts and so does the associative desorption energy of nitrogen, which is confirmed by the computational (first principle molecular simulation) investigation. This was corroborated with kinetic studies which were carried out experimentally and proposed that the recombination of nitrogen is the rate-limiting step for nickel catalysts [66]. Addendum to this, in the work by Ertl and Rustig [67], they calculated and found the activation energy of the decomposition of ammonia to be 197 kJ/mol and this was over clean a clean nickel surface.

Muroyama et al. [4] showed Ni/Al_2O_3 catalyst exhibited the highest activity of all their studied catalysts with Ni/La_2O_3 showing similar activity to that of Ni/Al_2O_3, though La_2O_3 has its surface area much smaller compared to that of Al_2O_3. Studies showed that ammonia decomposition is promoted by the rear earth metals, i.e., La and Ce [62, 68, 69].

Chellappa et al. [29] reported the kinetic data for a Ni-based catalyst for NH_3 decomposition at pressures above 100 Torr and temperatures greater than

793 K. It is a first-order reaction with a rate constant, k, of 1.309×10^{12} $\exp(-49{,}229/RT)/(g_{cat}\ h)$.

Deng et al. [70] reported high activity with high surface area $Ce_{0.8}Zr_{0.2}O_2$ solid solution supported Ni catalysts for NH_3 decomposition where 95.7% NH_3 conversion at 550 °C was achieved for this catalyst. Cao et al. [71] showed Ni/MRM catalysts to have very high activity in catalytic ammonia decomposition, with the 12 wt% Ni/MRM catalyst that was calcined at 600 °C displaying the highest activity having 95.5% conversion at 700 °C at 30,000 h^{-1} GHSV.

Even though substantial efforts have been directed toward the development and design of nickel in recent time, satisfactory investigations of many support materials for nickel catalysts are yet to be done.

The activity of nickel-based catalysts for NH_3 decomposition is greatly affected by some factors which include nickel particle size, phase formation, and nickel loading. In the study of reaction kinetics by Zhang et al. [72], a significant influence of Ni particle size is on the turnover frequency (TOF). They cited that a decrease in the average size of the nickel particle from 4 to 2 nm brought about tenfold increase of TOF. The TOF relationship with the nickel particle size is shown in Fig. 8, revealed that nickel particles having an average size lesser than 2.9 nm with the peak value of 2.3 nm is highly active for NH_3 decomposition [62].

The methods of preparation of nickel catalysts have also been reported to play a role in their activity. Co-precipitation methods had been found to give more active catalysts than the impregnation method due to the variation in nickel particle size and metal (nickel) dispersion [62, 72].

The nickel loading is another influencing factor that impacts the reactivity. The loading effects are felt on the nickel particle size as well as the formed phase. The trend is shown in Fig. 9 for nickel loading on Al_2O_3 support. At low loading, high coverage of Ni over the support is achieved which are majorly α-NiO phase while γ-Ni is formed at high loading and this phase formation affects their activity. The

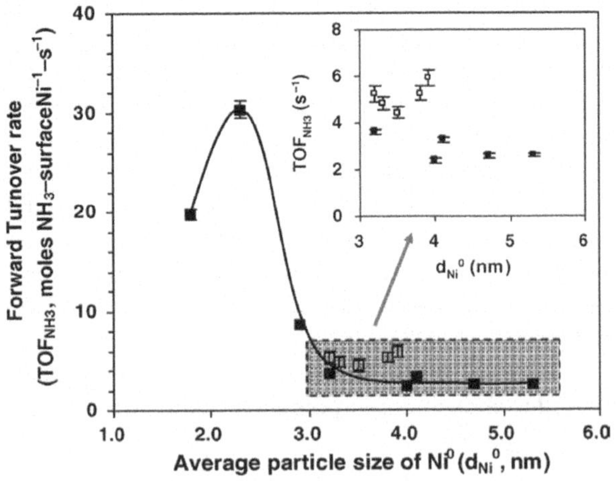

Fig. 8 Relationship between the forward turnover frequency (measured activity) versus the average particle size of Ni^0 where the hollows: Ni/La–Al_2O_3; solids: Ni/Al_2O_3. The inserted graph is an expansion of the gray color [62]

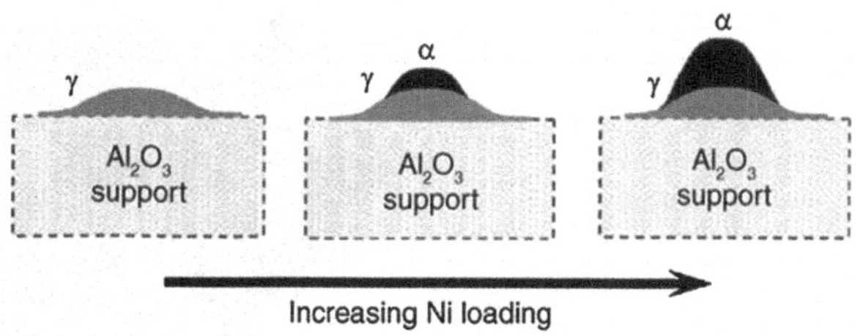

Fig. 9 Ni supported on Al_2O_3 interaction convention [72]

latter phase brings about a decrease in the surface area and as a result decreases in activity [72].

Even though enormous research efforts have been expended, irreducible nickel compound formation, the high sensitivity of nickel structure and high binding energy of nickel active sites with hydrogen are pending matters that need further research. Likewise, the physicochemical properties' effect on the rate of reaction and applicable elementary steps are yet to be well established, and this has resulted in many conflicting proposals which are still been put forward by researchers.

5.4 Bimetallic (Transition Metal) Catalysts

Recently, bimetallic catalysts are turning to be a novel set of materials in the field of heterogeneous catalysis due to their increased activity and selectivity coupled with enhanced stability which cannot be matched up by their corresponding monometallic catalysts, and this has been reported to be due to the synergetic effect of the metal combination [73, 74].

The volcano-type interaction of the nitrogen atom binding energy on the transition metal surface with the activity of ammonia decomposition has stimulated the development and design of different bimetallic catalysts which have a similar activity or superior to that of ruthenium-based catalysts [75, 76]. However, the nature of the active sites and cause of the enhanced activity are yet to be completely understood, e.g., if the loading of the guest metal is $\approx 50\%$, the bimetallic surface would have improved activity. Likewise, it was shown by the combined scanning tunnel microscopy and density functional theory (DFT) studies that isolated metal atoms, defect, and surface microstructure can significantly affect the ammonia decomposition reaction [74].

Depending on the distribution of the two metals, the classification of the bimetallic systems are of different kinds. But basically, two of them will be discussed below: alloys and monolayer bimetallic.

An alloy is formed when there is an even distribution of atoms both within the core and surface. There is a pure phase composition of the two combined metals through the bulk and surface [77]. Figure 10 shows a diagrammatic representation of a bimetallic alloy structure.

In this type of bimetallic structure system, the bimetallic surface properties are assumed to be an approximation of the linear combination of the mother metals

Fig. 10 A bimetallic alloy representation with uniform atom distribution [78]

and the prediction of this kind of active catalyst structure can be done by periodic interpolation method which involve the mixing of metals with low binding energy with one of high binding energy of nitrogen to get a bimetallic of intermediate binding energy for this cracking process [79].

The Co–Mo bimetallic catalysts are examples of this kind of structural system, with Co and Mo with low and high binding energies for nitrogen, respectively. They are also very active in this decomposition reaction and [80]. Figure 11 shows the volcano-type relationship of the turnover frequency (TOF) which measures the catalytic activity of different monometallic catalysts and Co–Mo for ammonia synthesis and their individual nitrogen binding energy. From the figure, we can see that Co–Mo bimetallic catalyst has the highest activity for the decomposition of ammonia, and this has made a lot of experimental works to focus on it for ammonia decomposition and Co_3Mo_3N was claimed to be the active phase which was created on the surface during the reaction process [35, 78]. The effect of supports for Co–Mo system for

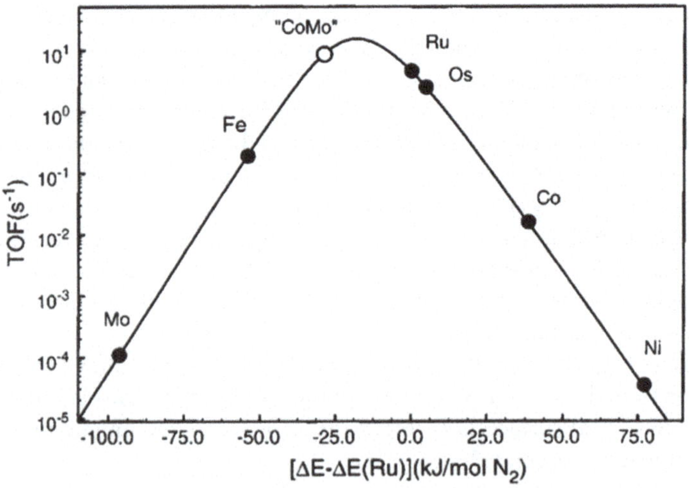

Fig. 11 Turnover frequency versus nitrogen binding energy [79]

NH_3 decomposition has been reported to be in the following descending order of activity γ-Al_2O_3 > MCM-41 > SiO_2 [78].

More often, there is a segregation of one metal of the metals to the surface for many bimetallic alloys owing to adsorbate-induced configuration or due to alloyed system surface energy minimization and have different surface properties to that of the perfectly alloyed surface due to the segregation. The segregation thus invalidates the periodic interpolation prediction method to create a bimetallic catalyst with desired properties [78].

As for the monolayer bimetallic system structure, it contains a monolayer of an admetal in the upper layers of a host metal. A core–shell bimetallic (one metal is encased on the surface by the monolayer of the second one) can be model from these surfaces, and likewise, they can be used for the representation of the segregated surface of an alloy. A surface configuration can be formed when the admetal is over the host metal while a subsurface configuration is formed when the admetal is below, as shown in Fig. 12. It has been shown both by DFT calculation and experimentally that these two configurations have very different properties from themselves and from the parent metals [78]. Depending on the configuration, the binding energies of the monolayer bimetallic surface are higher or lower than the parent metals and they do not depend on the binding energies of the parent metals. This nonlinear characteristic has also hindered the rational design of this unique catalyst [81].

The high stability of bimetallic catalysts during the reaction and their segregation to monometallic phase will change the surface properties and subsequently affects their catalytic activity [75]. Also, the thermal treatment conditions or high reaction temperatures which enhance the formation of less reducible oxides like Co–Si, Ni–Al, Co–Al, and Fe–Al when supported on silica or alumina, can consequence in a lower catalytic activity [82].

Like Mo–Co catalyst, Fe–Co is another potential bimetallic catalyst for ammonia decomposition which has shown experimentally to be very active when confined on the internal surface of CNT with the rate of formation of hydrogen being 4080 mol/mol h at 600 °C [82]. It is important to confine the Fe–Co on the internal surface of CNT to prevent it from sintering, and this also enhances its stability for more than 1000 h of operation.

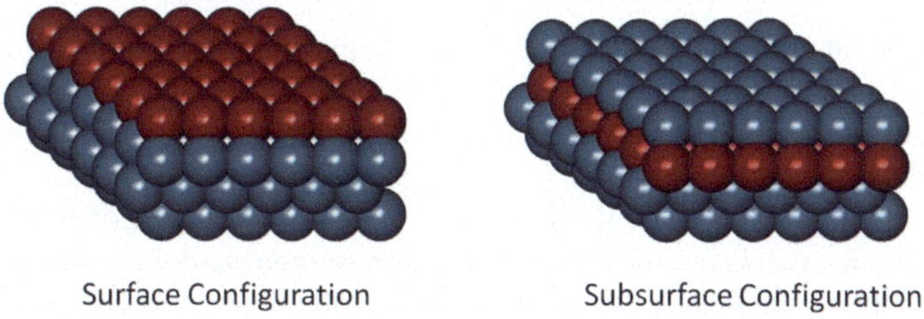

Fig. 12 A core–shell (monolayer) bimetallic representation with the colors indicating different metals [78]

Also, other supports have been reported to possess positive effects on the activity of the bimetallic systems, for example, NiFe/Al_2O_3 gives a higher hydrogen formation rate (640 mol/mol h) than Fe–Co (79 mol/mol h) at low temperature 500 °C. Also, NiFe showed the highest activity having alumina and/or Mg–Al-spinel as support compared to TiO_2, ZrO_2, and SiO_2, and this has been attributed to the latter supports surface area loss due to reduction at 800 °C [78].

A lot of work is being done to design and develop different bimetallic catalysts with the hope that it will give an optimum and economic solution to the catalytic decomposition of ammonia.

5.5 Transition Metal Nitrides

The number of accessible nitride materials has increased in the late century due to the advancement in preparation methods [83], e.g., the nitrides with spinel structure were recently prepared for the first time [84] even though oxynitrides of the same structure have been reported earlier.

Transition metal nitrides/carbides have been the focus of much attention due to two major themes:

(i) Good documentation of the ability of some transition metal nitrides to demonstrate good catalytic properties for numerous reactions due to their electronic (DOS) similarities with the Group 8 noble metals [77, 83, 85, 86].
(ii) The acid–base characteristics of nitrides and oxynitrides.

There are three subdivisions of nitrides:

(i) Ionic (salt-like) nitrides: They are formed from electropositive elements like Ca, Ba, Mg, Cu, Li, and Zn, and they have a formula based on the combination of N^{3-} ions with metal ions.
(ii) Covalent nitrides: They are formed by less electropositive elements like Si, P, S, and C.
(iii) Interstitial nitrides: They are formed by the transition metals having their interstitial sites occupied by nitrogen in close-packed metallic structures.

The nitrides are also classified based on the number of metals in the main structure as:

(i) Binary (with one metal); (ii) ternary (with two metals); (iii) quaternary (with three metals).

Of all these classifications, the interstitial nitrides have been of high attention in heterogeneous catalysis, most especially due to their acid–base features. From all these catalysts, the most well deeply studied is the binary molybdenum nitrides. Among the three different phases (γ-, β-, and δ-phases), the catalytic behavior of γ-molybdenum nitride has been the most widely reported [83, 87]. The best description for the γ-Mo_2N is a face-centered cubic arrangement of Mo atoms with a half octahedral interstice being occupied by nitrogen atom as shown in Fig. 13.

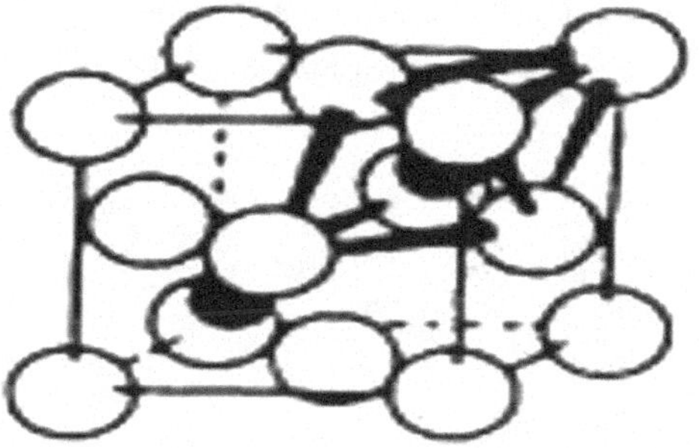

Fig. 13 γ-Mo_2N unit cell; molybdenum—white, nitrogen—black [88]

Zheng [59] studied and compared the activity of molybdenum nitride and carbide for NH_3 decomposition and observed that the nitrogen catalyst interaction which happened over the step locations and terrace of the Mo_2C and MoN are the preferred active sites for this dissociation reaction. The nitrogen atom is more tightly bound to the step edge on MoN than that of Mo_2C, and Mo_2C has the advantage of N–N recombination by surface diffusion.

Just in recent years, there are reports on ternary nitrides with Co_3Mo_3N being the most widely reported [88]. Duan et al. [89] reported Mo–Co bimetallic catalyst and reported that during ammonia decomposition reaction the Co_3Mo_3N phase was formed which is responsible for high activity of the catalyst for this reaction.

As for ammonia synthesis and decomposition, the interactions between metals and nitrogen are of great significance since the rate-determining steps for the two reactions are absorption and desorption of nitrogen, respectively [57].

Molybdenum- and vanadium-based transition metal nitrides have been tested for ammonia decomposition, and the high activity and low cost of molybdenum nitride have attracted the development of molybdenum nitride-based catalysts. There are other reports as well that suggested that the phases of Mo_2N, Ni_2Mo_3N, Ni_3Mo_3N, and Co_3Mo_3N are the active species for the ammonia decomposition reaction [90].

For ammonia decomposition, there are reports that suggest that Mo_2N is promoted by the addition of Co, Ni, and Fe species and showed higher activity compared to the base catalyst. Jolaoso et al. [91] concluded the higher yield by Co_3Mo_3N was due to the synergistic effect of Co and Mo by the decrease in the N-adsorption energy by Co over Mo_2N, and Srifa et al. [90] reported the same. From all the above-mentioned species, Zaman et al. [92] reported the highest activity, >95% conversion at 6000 h^{-1}, GHSV for ammonia decomposition over Ni_2Mo_3N at 525 °C making Ni to be the best structural promoter reported so far for Mo_2N and also mentioned that the high yield was also due to their method of preparation (sol–gel chelatization), in attaining higher surface area, of the catalyst. Also, in another report by Zaman et al. [93] showed that Fe_3Mo_3N is not a good catalyst for this reaction (Table 4).

Also, Zaman [96] reported a DFT simulation over Mo_3N_2 cluster (modeling Mo_2N catalyst) for ammonia decomposition and found that the abstraction of the second hydrogen from ammonia is the rate-limiting step according to the following surface elementary reaction ($NH_2{*}_{ad} \rightarrow NH{*}_{ad} + H{*}_{ad}$) similar to Pt(111) and Ni(110) plane.

Table 4 Catalytic activity of different nitride-based catalysts for ammonia decomposition

Catalyst	Temperature (°C)	GHSV (h^{-1})	Conversion (%)	Activation energy, E_a (kJ mol^{-1})	Reference
Mo_2N	550	6000	71.9	131.30	[91]
Co_3Mo_3N	550	6000	97.0	60.11	[91]
Fe_3Mo_3N	550	6000	78.2	72.88	[93]
Ni_2Mo_3N	550	6000	99.0	66.10	[92]
Ni_3Mo_3N	550	6000	83.0	83.80	[90]
Ni_2Mo_3N	550	30,000	84.0	66.10	[92]
NiMoN/α-Al_2O_3	600	3600	79.0	–	[94]
Ni_2Mo_3N	500	21,600	29.0	–	[95]

The energy barrier for nitrogen dissociation is 35.19 kcal/mol on the Mo_3N_2, much higher compared to ruthenium and vanadium catalysts, which makes Mo_2N much less active catalyst compared to Ru catalysts. Later on, Srifa et al. [90] suggest that incorporation of Ni and Co reduces the nitrogen recombination energy for Mo_2N catalysts and thus shows better activity compared to Mo_2N.

With all these been done, more works are still required for the mechanism of formation of different metal nitrides through different preparation methods for stability and activity optimization for both ammonia decomposition and other reactions.

6 Conclusions

Ammonia decomposition to generate CO_x-free hydrogen is a quite viable and environmentally friendly solution to the onboard hydrogen generation for fuel cell-driven vehicles. The challenge is to generate hydrogen by cracking ammonia at low temperature, i.e., less than 500 °C. Unfortunately, except the precious noble metal ruthenium, none of the catalysts showed 100% conversion at high GHSV (>30,000 h^{-1}) at 500 °C. This chapter has discussed several alternative options compared to ruthenium-based catalysts that have been investigated so far by different researchers. Of them, ternary nitrides (Co_3Mo_3N and Ni_2Mo_3N) have shown permissible potential for ammonia decomposition at low temperature compared to other nonprecious reported catalysts. In the near future, more research based on these ternary nitride-based catalysts will be available and a viable solution for ammonia decomposition will be achievable.

Acknowledgements Authors gratefully thank for the support provided by the Chemical and Materials Engineering Department, Faculty of Engineering, King Abdulaziz University, Jeddah, Saudi Arabia. Special thanks to Professor Lachezar Angelov Petrov, SABIC Chair of Catalysis at King Abdulaziz University, for his suppot to allow Dr. Sharif F. Zaman to work in the field of ammonia decompostion reaction.

References

1. Doelle M (2004) Climate change and human rights: the role of international human rights in motivating states to take climate change seriously. Macquarie J Int Comp Environ Law 1:179
2. IPCC fourth assessment report (2007)
3. High human cost of weather-related disasters detailed in report, News and Media, United Nations Radio (2015)
4. Muroyama H, Saburi C, Matsui T, Eguchi K (2012) Ammonia decomposition over Ni/La_2O_3 catalyst for on-site generation of hydrogen. Appl Catal A: Gen 443–444:119–124. https://doi.org/10.1016/j.apcata.2012.07.031
5. William IFD, Joshua WM, Samantha KC, Hazel MAH, James DT, Thomas JW, Martin OJ (2014) Hydrogen production from ammonia using sodium amide. J Am Chem Soc 136:13082–13085. https://doi.org/10.1021/ja5042836
6. Yin SF, Xu BQ, Zhou XP, Au CT (2004) A mini-review on ammonia decomposition catalysts for on-site generation of hydrogen for fuel cell applications. Appl Catal A 277:1–9. https://doi.org/10.1016/j.apcata.2004.09.020
7. Léon A (ed) (2008) Hydrogen technology: mobile and portable applications. Springer series in green energy and technology. Springer
8. Web source: https://energy.gov/eere/energybasics/articles/hydrogen-and-fuel-cell-technology-basics
9. Web source: www.HydrogenAssociation.org
10. Cheddie D (2012) Ammonia as a hydrogen source for fuel cells: a review. In: Minic D (ed) Hydrogen energy—challenges and perspectives. InTech. https://doi.org/10.5772/47759
11. Avery WH (1988) A role for ammonia in the hydrogen economy. Int J Hydrogen Energy 13:761–773. https://doi.org/10.1016/0360-3199(88)90037-7
12. Christensen CH, Johannessen T, Sorensen RZ, Norskov JK (2006) Towards an ammonia-mediated hydrogen economy. Catal Today 111:140–144. https://doi.org/10.1016/j.cattod.2005.10.011
13. Jensen JO, Vestbo AP, Li Q, Bjerrum NJ (2007) The energy efficiency of onboard hydrogen storage. J Alloys Compd 446–447:723–728. https://doi.org/10.1016/j.jallcom.2007.04.051
14. Web source: United States geological survey publication. https://minerals.usgs.gov/minerals/pubs/commodity/nitrogen/mcs-2011-nitro.pdf
15. Web source: United States geological survey publication: https://minerals.usgs.gov/minerals/pubs/commodity/nitrogen/mcs-2017-nitro.pdf
16. Web source: http://www.roperld.com/science/minerals/ammonia.htm
17. Web source: http://www.greener-industry.org.uk/pages/ammonia/1ammoniaapq.htm
18. Clausen JF, Zee CA, TRW-Systems and Energy, EPA Contract 68-02-2635
19. International Energy Agency (IEA) (2007) Key world energy statistics
20. B.P. Statistical review of world energy, June 2007.
21. Schlogl R (2003) Catalytic synthesis of ammonia—a "never-ending story"? Angew Chem Int Ed 42:2004–2008. https://doi.org/10.1002/anie.200301553
22. Appl M (2007) Ullmann's encyclopedia of industrial chemistry: ammonia. Wiley-VCH Verlag GmbH & Co. KGaA, Weinheim
23. Kaye IW, Bloomfield DP (1998) Portable ammonia powered fuel cell. In: Conference of power source, pp 408–409
24. Lipman T, Shah N (2007). UC Berkeley Transportation Sustainability Research Center, UC Berkeley. Retrieved from: http://escholarship.org/uc/item/7z69v4wp
25. White AH, Melville WM (1905) The decomposition of ammonia at high temperatures. J Am Chem Soc 27:373–386. https://doi.org/10.1021/ja01982a005
26. Haber F, Van Oordt G (1904) Z Anorg Chem 43:111; 44:341 (1905); Topham SA (1985) The history of the catalytic synthesis of ammonia. In: Anderson JR et al (eds) Catalysis. Springer, Berlin, Heidelberg. https://doi.org/10.1007/978-3-642-93281-6_1
27. Hansen JB (1995) In: Nielsen A (ed) Ammonia, catalysis and manufacture. Springer, Heidelberg, pp 149–198

28. Tsai W, Weinberg WH (1987) Steady-state decomposition of ammonia on the ruthenium (001) surface. J Phys Chem 91:5307. https://doi.org/10.1021/j100304a034
29. Chellappa AS, Fischer CM, Thomson WJ (2002) Ammonia decomposition kinetics over Ni-Pt/Al_2O_3 for PEM fuel cell applications. Appl Catal A: Gen 227:231–240. https://doi.org/10.1016/S0926-860X(01)00941-3
30. Green L (1982) An ammonia energy vector for hydrogen economy. Int J Hydrogen Energy 7:355–359. https://doi.org/10.1016/0360-3199(82)90128-8
31. Bell TE, Torrent-Murciano L (2016) H_2 production via ammonia decomposition using non-Noble metal catalysts: a review. Top Catal 59:1438–1457. https://doi.org/10.1007/s11244-016-0653-4
32. Georgeta P, Aline A (2011) The poisoning level of Pt/C catalysts used in PEM fuel cells by the hydrogen feed gas impurities: the bonding strength. Int J Hydrogen Energy 36:6817–6825. https://doi.org/10.1016/j.ijhydene.2011.03.018
33. Yin SF, Zhang QH, Xu BQ, Zhu WX, Ng CF, Au CT (2004) Investigation on the catalysis of CO_x-free hydrogen generation from ammonia. J Catal 224:384–396. https://doi.org/10.1016/j.jcat.2004.03.008
34. Klerke A, Christensen CH, Nørskov JK, Vegge T (2008) Ammonia for hydrogen storage: challenge and opportunities. J Mater Chem 18:2304–2310. https://doi.org/10.1039/B720020J
35. Podila S, Zaman SF, Driss H, Alhamed Y, Al-Zahrani AA, Petrov LA (2016) Hydrogen production by ammonia decomposition using high surface area Mo_2N and Co_3Mo_3N catalysts. Catal Sci Technol 6:1496–1506. https://doi.org/10.1039/C5CY00871A
36. Yin SF, Xu BQ, Wang SJ, Ng CF, Au CT (2004) Magnesia-carbon nanotubes (MgO–CNTs) nanocomposite: novel support of Ru catalyst for the generation of CO_x-free hydrogen from ammonia. Cat Lett 96:113–116. https://doi.org/10.1023/B:CATL.0000
37. Choudhary TV, Sivadinarayana C, Goodman DW (2001) Catalytic ammonia decomposition: CO_x-free hydrogen production for fuel cell applications. Catal Lett 72:197–201. https://doi.org/10.1023/A:1009023825549
38. Ganley JC, Thomas FS, Seebauer EG, Masel RI (2004) A priori catalytic activity correlations: the difficult case of hydrogen production from ammonia. Catal Lett 96(3–4):117–122. https://doi.org/10.1023/B:CATL.0000030108.50691.d4
39. Li L, Zhu ZH, Yan ZF, Lu GQ, Rintoul L (2007) Catalytic ammonia decomposition over Ru/carbon catalysts: the importance of the structure of carbon support. Appl Catal A: Gen 320:166–172. https://doi.org/10.1016/j.apcata.2007.01.029
40. Egawa C, Nishida T, Naito S, Tamaru K (1984) ammonia decomposition on (1 1 10) and (0 0 1) surfaces of ruthenium. J Chem Soc Faraday Trans 1: Phys Chem Condens Phases 80:1595–1604. https://doi.org/10.1039/f19848001595
41. Dietrich H, Jacobi K, Ertl G (1996) Decomposition of NH_3 on Ru(111). Surf Sci 352–354:138–141. https://doi.org/10.1016/0039-6028(95)01120-X
42. Zheng W, Zhang J, Zhu B, Blume R, Zhang Y, Schlichte K, Schlögl R, Schüth F, Su DS (2010) Structure–function correlation for Ru/CNT in the catalytic decomposition of ammonia. ChemSusChem 39:226–230. https://doi.org/10.1002/cssc.200900217
43. Sorensen RZ, Nielsen LJE, Jensen S, Hansen O, Johannesen T, Quaade U, Christensen CH (2005) Catalytic ammonia decomposition: miniaturized production of CO_x-free hydrogen for fuel cells. Catal Commun 6:229–232. https://doi.org/10.1016/j.catcom.2005.01.005
44. Wang Z, Cai Z, Wei Z (2019) Highly active ruthenium catalyst supported on barium hexaaluminate for ammonia decomposition to CO_x-free hydrogen. ACS Sustain Chem Eng 7(9):8226–8235. https://doi.org/10.1021/acssuschemeng.8b06308
45. (a) Robertson AJB (1975) The early history of catalysis. Platin Met Rev 19:64–69. (b) Amano A, Taylor H (1954) The decomposition of ammonia on ruthenium, rhodium and palladium catalysts supported on alumina. J Am Chem Soc 76:4201–4204. https://doi.org/10.1021/ja01645a057
46. Tanaka KI, Tamaru K (1966) On one general principle of the catalytic activity of metals. Kinetikai Kataliz 7:242–247

47. Papapolymerou G, Bontozoglou V (1997) Decomposition of NH_3 on Pd and Ir comparison with Pt and Rh. J Mol Catal A: Chem 120:165–171. https://doi.org/10.1016/S1381-1169(96)00428-1
48. Zhang J, Comotti M, Schuthe F, Schlogl R, Su DS (2007) Commercial Fe- or Co-containing carbon nanotubes as catalysts for NH_3 decomposition. Chem Commun 19:1916–1918. https://doi.org/10.1039/B700969K
49. Lendzion-Bielun Z, Pelka R, Arabczyk W (2009) Study of the kinetics of ammonia synthesis and decomposition on iron and cobalt catalysts. Catal Lett 129:119–121. https://doi.org/10.1007/s10562-008-9785-x
50. Zhang H, Alhamed YA, Al-Zahrani AA, Daous MA, Inokawa H, Kojima Y, Petrov LA (2014) Tuning catalytic performances of cobalt catalysts for clean hydrogen generation via variation of the type of carbon support and catalyst post-treatment temperature. Int J Hydrogen Energy 39:17573–17582. https://doi.org/10.1016/j.ijhydene.2014.07.183
51. Zhang J, Xu HY, Jin XL, Ge QJ, Li WZ (2006) Highly efficient Ru/MgO catalysts for NH_3 decomposition: synthesis, characterization and promoter effect. Catal Commun 7:148–152. https://doi.org/10.1016/j.catcom.2005.10.002
52. Zhao C, Yang Y, Wu Z, Fiel M, Fang XJ (2014) Synthesis and facile size control of well-dispersed cobalt nanoparticles supported on ordered mesoporous carbon. J Mater Chem A 2:19903–19913. https://doi.org/10.1039/C4TA04561K
53. Podila S, Alhamed YA, AlZahrani AA, Petrov LA (2015) Hydrogen production by ammonia decomposition using Co catalyst supported on Mg mixed oxide systems. Int J Hydrogen Energy 40:15411–15422. https://doi.org/10.1016/j.ijhydene.2015.09.057
54. Gu YQ, Jin Z, Zhang H, Xu RJ, Zheng MJ, Guo YM, Song QS, Jia CJ (2015) Transition metal nanoparticles dispersed in an alumina matrix as active and stable catalysts for CO_x-free hydrogen production from ammonia. J Mater Chem A 3:17172–17180. https://doi.org/10.1039/C5TA04179A
55. Love KS, Emmett PH (1941) The catalytic decomposition of ammonia over iron synthetic ammonia catalysts. J Am Chem Soc 63:3297–3308. https://doi.org/10.1021/ja01857a019
56. Dannstadt S (2000) Ullmann's encyclopedia of industrial chemistry. Wiley, Weinheim
57. Schlögl R (2008) Ammonia synthesis. In: Ertl G et al (eds) Handbook of heterogeneous catalysis. Wiley
58. Mittasch A, Frankenburg W (1950). Early studies of multicomponent catalysts. In: Komarewsky VI, Frankenburg WG, Rideal EK (eds) Advances in catalysis. Academic Press, pp 81–104. https://doi.org/10.1016/s0360-0564(08)60375-2
59. (a) Zheng W (2011) Nanomaterials for ammonia decomposition. Ph.D. thesis, Universitat Berlin. (b) Zheng W, Cotter TP, Kaghazchi P, Jacob T, Frank B, Schlichte K, Zhang W, Su DS, Schüth F, Schlögl R (2013) Experimental and theoretical investigation of molybdenum carbide and nitride as catalysts for ammonia decomposition. J Am Chem Soc 135:3458–3464. https://doi.org/10.1021/ja309734u
60. Arabczyk W, Zamlynny J (1999) Study of the ammonia decomposition over iron catalysts. Catal Lett 60:167–171. https://doi.org/10.1023/A:101900702
61. Kielbasa K, Pelka R, Arabczyk W (2010) Studies of the kinetics of ammonia decomposition on promoted nanocrystalline iron using gas phases of different nitriding degree. J Phys Chem A 114:4531–4534. https://doi.org/10.1021/jp9099286
62. Zhang J, Xu H, Li W (2005) Kinetic study of NH_3 decomposition over Ni nanoparticles: the role of La promoter, structure sensitivity and compensation effect. Appl Catal A 296:257–267. https://doi.org/10.1016/j.apcata.2005.08.046
63. Tsai W, Vajo JJ, Weinberg WH (1985) Inhibition by hydrogen of the heterogeneous decomposition of ammonia on platinum. J Phys Chem 89:4926–4932. https://doi.org/10.1021/j100269a009
64. Roy SK, Ray N, Mukheriee D (1975) Kinetics and mechanism of ammonia decomposition over alumina supported nickel catalysts. Plan Dev Div 41:485–495
65. Okura K, Miyazaki K, Muroyama H, Matsui T, Eguchi K (2018) Ammonia decomposition over Ni catalysts supported on perovskite-type oxides for the on-site generation of hydrogen. RSC Adv. 8:32102–32110. https://doi.org/10.1039/C8RA06100A

66. Duan X, Qian G, Liu Y, Ji J, Zhou X, Chen D, Yuan W (2013) Structure sensitivity of ammonia decomposition over Ni catalysts: a computational and experimental study. Fuel Process Technol 108:112–117. https://doi.org/10.1016/j.fuproc.2012.05.030
67. Ertl G, Rustig J (1982) Decomposition of NH_3 on nickel: absence of a magneto-catalytic effect. Surf Sci 119:314–318. https://doi.org/10.1016/0039-6028(82)90173-X
68. McCabe RW (1983) Kinetics of ammonia decomposition on nickel. J Catal 79:445–450. https://doi.org/10.1016/0021-9517(83)90337-8
69. Liu H, Wang H, Shen J, Sun Y, Liu Z (2008) Promotion effect of cerium and lanthanum oxides on Ni/SBA-15 catalyst for ammonia decomposition. Catal Today 131:444–449. https://doi.org/10.1016/j.cattod.2007.10.048
70. Deng QF, Zhang H, Hou XX, Ren TZ, Yunan ZY (2012) High-surface-area $Ce_{0.8}Zr_{0.2}O_2$ solid solutions supported catalysts for ammonia decomposition to hydrogen. Int J Hydrogen Energy 37:15901–15907. https://doi.org/10.1016/j.ijhydene.2012.08.069
71. Cao JL, Yan ZL, Deng QF, Wang Y, Yuan ZY, Sun G, Jia TK, Wang XD, Bala H, Zhang ZY (2015) Mesoporous modified-red-mud supported Ni catalysts for ammonia decomposition to hydrogen. Int J Hydrogen Energy 39:15411–15422. https://doi.org/10.1016/j.ijhydene.2014.01.169
72. Zhang J, Xu H, Jin X, Ge Q, Li W (2005) Characterizations and activities of the nano-sized Ni/Al_2O_3 and $Ni/La–Al_2O_3$ catalysts for NH_3 decomposition. Appl Catal A: Gen 290:87–96. https://doi.org/10.1016/j.apcata.2005.05.020
73. Web source: https://www.strem.com/uploads/resources/documents/electroless_deposition_for_the_synthesis_of_bimetallic_catalysts.pdf
74. Guo W, Vlachos DG (2015) Patched bimetallic surfaces are active catalysts for ammonia decomposition. Nat Commun 6:8619–8626. https://doi.org/10.1038/ncomms9619
75. Hansgen DA, Vlachos DG, Chen JG (2010) Using first principles to predict bimetallic catalysts for the ammonia decomposition reaction. Nat Chem 2:484–489. https://doi.org/10.1038/nchem.626
76. Boisen A, Dahl S, Norskov JK, Christensen CH (2005) Why the optimal ammonia synthesis catalyst is not the optimal for ammonia decomposition. J Catal 230:309–312. https://doi.org/10.1016/j.jcat.2004.12.013
77. Leclercq L, Provost M, Pastor H, Grimblot J, Hardy AM, Gendembre L, Leclercq G (1989) Catalytic properties of transition metal carbides: I. Preparation and physical characterization of bulk mixed carbides of molybdenum and tungsten. J Catal 117:371–383. https://doi.org/10.1016/0021-9517(89)90348-5
78. Hansgen DA (2011) Rational catalyst design for the ammonia decomposition reaction. Ph.D. thesis
79. Jacobsen CJH, Dahl S, Clausen BS, Bahn S, Logadottir A, Norskov JK (2001) Catalyst design by interpolation in the periodic table: bimetallic-ammonia synthesis catalysts. J Am Chem Soc 123:8404–8405. https://doi.org/10.1021/ja010963d
80. Duan X, Ji J, Yan X, Qian G, Chen D, Zhou X (2016) Understanding Co–Mo catalyzed ammonia decomposition: influence of calcination atmosphere and identification of active phase. ChemCatChem 8:938–945. https://doi.org/10.1002/cctc.201501275
81. Chen JG, Menning CA, Zellner MB (2008) Monolayer bimetallic surfaces: experimental and theoretical studies of trends in electronic and chemical properties. Surf Sci Rep 63:201–254. https://doi.org/10.1016/j.surfrep.2008.02.001
82. Zhang J, Muller JO, Zheng W, Wang D, Su D, Schlogl R (2008) Individual Fe–Co alloy nanoparticles on carbon nanotubes: structural and catalytic properties. Nano Lett 8(9):2738–2743. https://doi.org/10.1021/nl8011984
83. Oyama ST (1996) In: Oyama ST (ed) The chemistry of transition metal carbides and nitrides. Blackie Academic and Professional, Glasgow. https://doi.org/10.1007/978-94-009-1565-7_1
84. Scotti N, Kockelmann W, Senker J, Traßel S, Jacobs H (1999) Sn3N4, a tin (IV) nitride—syntheses and the first crystal structure determination of a binary tin–nitrogen compound. Z Anorg Allg Chem 625:1435–1439. https://doi.org/10.1002/chin.199947033

85. Petterson PM, Das TK, Davis BH (2003) Carbon monoxide hydrogenation over molybdenum and tungsten carbides. Appl Catal A 251:449–455. https://doi.org/10.1016/S0926-860X(03)00371-5
86. Claridge JB, York APE, Brungs AJ, Marquez-Alvarez C, Sloan J, Tsang SC, Green MLH (1998) New catalysts for conversion of methane to syngas: molybdenum and tungsten carbide. J Catal 180:85–100. https://doi.org/10.1006/jcat.1998.2260
87. Marchand R, Gouin X, Tessier F, Laurent Y (1996) In: Oyama ST (ed) The chemistry of transition metal carbides and nitrides. Blackie Academic and Professional, Glasgow, p 252, chapter 13
88. Mckay D (2008) Catalysis over molybdenum containing nitride materials. Ph.D. thesis, University of Glasgow. http://theses.gla.ac.uk/id/eprint/174
89. Duan X, Qian G, Zhou X, Chen D, Yuan W (2012) MCM-41 supported CoMo bimetallic catalysts for enhanced hydrogen production by ammonia decomposition. J Chem Eng 207–208:103–108. https://doi.org/10.1016/j.cej.2012.05.100
90. Srifa A, Okura K, Okanishi T, Muroyama H, Matsui T, Eguchi K (2016) CO_x-free hydrogen production via ammonia decomposition over molybdenum nitride-based catalysts. Catal Sci Technol 6:7495–7504. https://doi.org/10.1039/C6CY01566B
91. Jolaoso LA, Zaman SF, Podila S, Driss H, Al-Zahrani AA, Daous MA, Petrov L (2018) Ammonia decomposition over citric acid induced γ-Mo_2N and Co_3Mo_3N catalysts. Int J Hydrogen Energy 43:4839–4844. https://doi.org/10.1016/j.ijhydene.2018.01.092
92. Zaman SF, Jolaoso LA, Podila S, Al-Zahrani AA, Alhamed YA, Daous HA, Petrov L (2018) Ammonia decomposition over citric acid chelated γ-Mo_2N and Ni_2Mo_3N catalysts. Int J Hydrogen Energy 43:17252–17258. https://doi.org/10.1016/j.ijhydene.2018.07.085
93. Zaman SF, Jolaoso LA, Al-Zahrani AA, Alhamed YA, Podila S, Driss H, Daous MA, Petrov L (2018) Study of Fe_3Mo_3N catalyst for ammonia decomposition. Bulg Chem Commun 50(Special issue H):181–188
94. Liang C, Li W, Wei Z, Xin Q, Li C (2000) Catalytic decomposition of ammonia over nitrided MoN_x/α-Al_2O_3 and $NiMoNy/\alpha$-Al_2O_3 catalysts. Ind Eng Chem Res 39:3694–3697. https://doi.org/10.1021/ie990931n
95. Leybo DV, Baiguzhina AN, Muratov DS, Arkhipov DI, Koles-nikov EA, Levina VV, Kosova NI, Kuznetsov DV (2016) Effects of composition and production route on structure and catalytic activity for ammonia decomposition reaction of ternary Ni–Mo nitride catalysts. Int J Hydrogen Energy 41:3854–3860. https://doi.org/10.1016/j.ijhydene.2015.12.171
96. Zaman SF (2018) A DFT study of ammonia dissociation over Mo_3N_2 cluster. Bulg Chem Commun 50(Special issue H):201–208

Electrocatalytic Nitrogen (N_2) Reduction

Chayanika Chaliha and Eeshan Kalita

Abstract Ammonia (NH_3) is an industrially important chemical for its use in manufacturing fertilizers, carbon-free fuel and synthesis of essential biological building blocks and as energy carrier. The most widely used industrial process NH_3 production, the Haber–Bosch process, has several bottlenecks such as high operational costs and high energy consumption and is a severe detriment for environment due to its large carbon footprint. In recent decades, electrocatalysis of N_2 to produce NH_3 has emerged as a sustainable alternative and provides an efficient means for the production of NH_3 from N_2 under ambient conditions. Till date, various kinds of electrocatalyst have been developed for N_2 reduction which covers a wide range of materials that includes noble metals, transition metals, single-atom catalyst and various carbon-based metal-free composites. Also, to increase the catalytic potential, different operational strategies have been developed that generate electrocatalysts with low overpotential. Molecular dynamics simulation-based studies have enabled the development of new generation electrocatalysts and have been investigated for their thermodynamics and mechanism in nitrogen reduction reaction (NRR). The combination of the theoretical and experimental provides a promising perspective to develop efficient electrocatalyst with increased surface active site, selectivity and durability in NRR.

Keywords Nitrogen · Ammonia · Electrocatalyst · NRR

1 Introduction

Ammonia (NH_3) is one of the most essential constituent chemicals for living beings as well as the earth's ecosystem as an essential biological building block. It is an indispensable precursor for fertilizers in agriculture production, a convenient hydrogen carrier and an emerging clean fuel [1]. Given the importance of ammonia, Nitrogen (N_2) fixation through the reduction of atmospheric nitrogen (N_2) to ammonia (NH_3)

C. Chaliha · E. Kalita (✉)
Department of Molecular Biology and Biotechnology, Tezpur Univeristy, Napaam, Assam 784028, India
e-mail: ekalita@tezu.ernet.in

Inamuddin et al. (eds.), *Sustainable Ammonia Production*,
Green Energy and Technology, https://doi.org/10.1007/978-3-030-35106-9_6

is considered to be one of the most important processes on earth [2]. In nature, N_2 fixation to ammonia is generally carried out by microorganisms catalyzed by the nitrogenase enzymes which contain iron-molybdenum cofactor (denoted as FeMo-co), and the required energy is delivered by ATP [3, 4]. In spite of the low energy and high electron efficiency of this method, the ultralow reaction rate restricts its use for practical application in large-scale ammonia synthesis [5]. However, the industrial N_2 reduction is primarily carried out through the Haber–Bosch process, wherein Fe/Ru-based catalyst processes nitrogen and hydrogen molecule at high pressure and temperature. The process operates at a cost of more than 1% of global energy consumption owing to the intense inertness of N_2 with N≡N bond energy of 940.95 kJ/mol accompanied with CO_2 emissions of ~2 ton, per ton of NH_3 produced [6]. Also, a capital expenditure of at least a million US dollars has to be invested during the construction of a Haber–Bosch chemical plant and is regarded as a major roadblock toward the establishment of fertilizer plants in developing countries. Thus, the energy consumption, environmental effect and production cost associated with Haber–Bosch process urge the development of an efficient, sustainable, cost effective and environment-friendly nitrogen fixation strategy that can be carried out with minimal infrastructure facility, low energy and at atmospheric pressure. Lately, various strategies have been explored, viz. biochemical, organometallic, photocatalytic and electrochemical approaches for the catalytic nitrogen reduction reaction (NRR) [7]. In recent times, electrocatalytic nitrogen reduction reactions have gained significant interest owing to the facile conditions during operation and renewable electric energy sources utilization. Also, electrocatalytic NRR facilitates low CO_2 emissions, as the proton source used herein is water instead of hydrogen gas used in Haber–Bosch process [8].

Fundamentally, in electrochemical NRR, ammonia synthesis is achieved via thermodynamic force controlled by applied voltage at room temperature and atmospheric pressure. The basic mechanism of electrochemical NRR involves a dissociative, an alternating associative and a distal associative pathway (Fig. 1) [9]. During the dissociative pathway, the triple bond of N_2 molecules is cleaved prior to the addition of any hydrogen. After cleavage, hydrogenation takes place via the alternating associative and a distal associative pathway. In the alternating pathway, hydrogenation takes place on the two N atoms with electron transfer coupled with proton, and at the final step, sequentially two NH_3 molecules get released. In the distal associative pathway, addition of hydrogen takes place in the distal N atom from the catalyst surface, and after the release of the first NH_3 molecule, the process repeats again [7].

2 Efficient NRR Electrocatalysts Design Strategies

The efficiency of a developed electrocatalyst is mediated by different factors related to its size, morphology, chemical and electronic structure. Tuning of these properties greatly enhances the electrocatalytic efficiency of the catalyst by lowering overpotential and increasing FE following which high yield of NH_3 is achieved. Various design

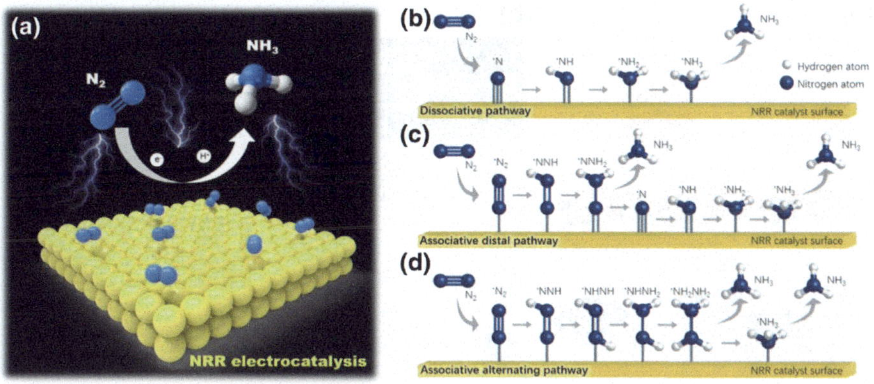

Fig. 1 Mechanism of Electrocatalytic nitrogen reduction reaction (NRR) **a**, Dissociative pathway **b**, Distal associative pathway **c**, Alternating associative pathway **d** (Reprinted from [10] with permission)

strategies of catalyst have been investigated with the main objective to increase the number of active sites and/or increase the catalytic activity of each surface active site (Fig. 2) [11]. Some of the efficient design strategies explored so far include the morphology and size engineering, crystal facet regulation and defect engineering via introduction of vacancy and doping that induces lattice distortion [12–14]. Insights into the mechanisms of these tunable design strategies to develop highly efficient NRR electrocatalyst are discussed in the following sections.

2.1 *Engineering Size and Morphology*

Modulating the size and morphology of electrochemical catalyst has been found to effectively enhance the catalytic property of the electrocatalyst [16]. Engineering of size and morphology result in tuning of the coordination sites, the binding strength of various reactants and the surface area with active sites. These properties greatly enhance the selectivity and durability of the electrocatalyst, thereby creating an impact on the catalytic performance of the electrocatalyst [17, 18]. For instance, hollow Au nanocages were found with higher electrocatalytic efficacy when compared to nanocubes, nanospheres and nanorods of Au. Here, the enhanced efficacy resulted due to the increased surface area with exposed active sites [19].

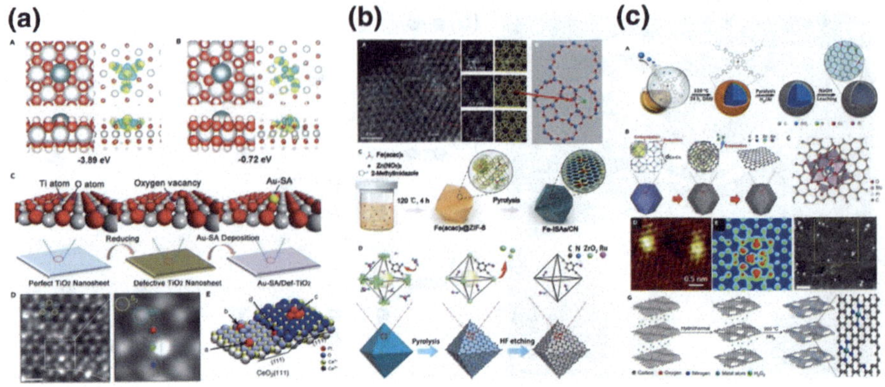

Fig. 2 **a** Defect Engineering Strategy for the Synthesis of SACs. (A and B) The most stable configuration and charge density difference of a single Pt atom trapped on $Ni(OH)_2$ nanoboards with abundant Ni^{2+} vacancies (A) and without Ni^{2+} vacancies (B) by DFT calculations. (C) The synthesis process for isolated Au atoms stabilized on defective TiO_2 nanosheets with oxygen vacancies. (D) HAADF-STEM image (left) and corresponding enlarged image (right) of single Co atoms anchored by S vacancies of MoS_2. (E) Schematic illustration of single Pt atoms located at the CeO_2 step edges, based on DFT calculations. **b** Spatial Confinement Strategy for the Synthesis of SACs. (A and B) AC-STEM images (A) and model (B) of single Pt atoms located in the zeolite KLTL. (C) Schematic preparation process of single-atom Fe catalyst. (D) Formation procedure of single Ru atoms immobilized on the carbonized skeleton of UiO-66-NH_2. **c** Coordination Design Strategy for the Synthesis of SACs. (A) Illustration of the synthesis of single Co atoms stabilized on hollow nitrogen-doped carbon. (B) Formation process of single Co atoms with precise N-coordination. (C) The most stable structure of single-atom Pt on PMA, based on DFT calculations. (D and E) Atomic-resolution scanning tunneling microscopy image (D) and the corresponding simulation image (E) of single Cu atoms embedded in nitrogen-doped graphene. (F) AC-STEM image of single Co atoms stabilized on nitrogen-doped graphene. (G) Schematic illustration of atomically dispersed transition metal (e.g., Fe, Co, or Ni) catalysts (Reprinted from [15] with permission)

2.2 Defect Engineering

Defect engineering tailors the electronic structure of the electrocatalyst, thereby modulating the catalytic performance. The introduction of artificial defects to modulate electrocatalytic performance includes vacancy of a lattice position due to missing of an atom, incorporation of foreign atoms (dopant) on the lattice of host materials and the strain on the surface of the electrocatalysts that alters the bond length and surface space of atom, thereby shaping the electronic structure [10, 20]. Vacancy engineering enables the transfer of charge between reactants and enhances adsorption energetics [21]. The vacancy engineering includes both anion and cation vacancies, wherein introduction of oxygen, cobalt and other metal atoms (Fe, Ni, W) has been widely explored [22, 23]. Introduction of O vacancy facilitates low activation energy whereas cation vacancy enhances the electrocatalytic property due to their orbital distribution of electron. The O vacancy induced by the removal of lattice oxygen atoms facilitates the adsorption of N_2 molecule by accumulating excess electrons through coordinately unsaturated metal atoms (CUS) [24, 25]. CUS metal sites are electron rich which acts as an active site, thereby promoting electron transfer between reactants in NRR [26]. Use of a dopant to artificially modulate electronic structure is a promising approach to increase the catalytic performance of electrocatalyst. The doped atoms are placed into the intermediate space in the lattice resulting in an enhancement in the activation of adsorbed molecules [27, 28]. Both metal and nonmetal atoms have been explored for their use as dopants that modulate the band gap resulting in low overpotential and high electrocatalytic efficiency [29]. For instance, carbon materials anchored to various nonmetals (S, N, P, B, F, etc.) and metals (Fe, Co, Ni, Cu, etc.) have been explored for their efficient use as catalysts in NRR [30, 31]. Further, in recent times, single metal atoms attached to the substrate materials is one form of doped composites wherein the doped atoms act as efficient catalytic sites modulating its electrocatalytic reduction of N_2 [32–34]. Hence, the defect engineering results in the development of microscopic strain that alters the electronic band structure, atomic bond length and spacing in surface, thereby effecting catalytic property.

2.3 *Crystal Facet Engineering*

Crystal facet engineering strategy mediates the atomic arrangement and coordination of a crystalline material employed for the exposure of facets that act as active sites for catalytic activity [35–37]. The exposed crystal facets greatly influence the enhancement of N_2 adsorption, thereby affecting the catalytic performance [38, 39]. Studies have shown that crystalline facets act as favorable active sites, and the intermediate reactants in NRR have strong binding efficiency toward these crystal facets as compared to flat facets, thereby enhancing electrocatalytic efficiency [40]. For instance, in Mo nanofilm, the diffraction plane at (1 10) was increased in four plane orientation which enhanced the catalytic efficiency to achieve NH_3 yield rate of 3.09×10^{-11} mol s^{-1} cm^{-2} with FE 0.72% at 0.49 V versus reversible hydrogen electrode (RHE) [41]. Owing to the strong adsorption on crystal facets, sulfate anions adsorption on the diffraction plane (1 1 1) of single-crystal Pt leads to a reduction in catalytic efficiency due to blocking of the active sites [20].

Till date, a large variety of electrocatalysts have been explored for NRR with various design strategies that include noble metal-based materials, transition metals and their composites, single-atom catalysts and carbon-based materials (Table 1). The electrocatalysts developed for NRR reactions and their catalytic efficiency are categorized and discussed in detail in the following sections.

2.4 *Noble Metals*

Noble metals have been explored for their application as electrocatalysts for N_2 reduction to ammonia owing to their good electrical conductivity. The morphology-controlled synthesis of noble metal to modulate geometric structures and electronic

Table 1 Summary of reported electrocatalyst for NRR

Material	Catalyst	NH_3 yield	FE (%)	Potential (V vs. RHE)	Refs.
Transition metal	ɣ-Fe_2O_3 nanoparticle	58.9 nmol h^{-1} cm^{-2}	1.96	0	[42]
	Fe_2O_3 nanorod	15.9 mg$h^{-1}$$mg^{-1}$	0.94	−0.8	[43]
	Mo_2N	1.99 μg mg^{-1} h^{-1}	42.3	−0.05	[44]
	MoS_2 nanosheet	8.08×10^{-11} mol $s^{-1}$$cm^{-1}$	1.17	−0.5	[45]
	MoS_2 nanoflower	29.28 μg h^{-1} mg^{-1}cat	8.34	−0.4	[46]
	Mn_3O_4 nanocube	11.6 μg h^{-1} mg^{-1}cat	3	−0.8	[47]
	Nb_2O_5 nanofiber	43.6 mgh^{-1} mg^{-1}cat	9.26	−0.55	[48]
	Cr_2O_3 microspheres	25.3 μgh–1 mg^{-1}cat	6.78	−0.9	[49]
	CeO_2 nanorod	16.4 μg h^{-1} mg^{-1}cat	3.7	−0.5	[50]
	THH Au nanorod	1.648 μg h^{-1} cm^{-2}	1.4	−0.2	[51]
	AuH nanocage	3.74 μg cm^{-2} h^{-1}	35.9	−0.4	[52]
	$Bi_4V_2O_{11}$–CeO_2	23.21 mg$h^{-1}$$mg^{-1}$cat	10.16	−0.2	[53]
	PdCu alloy	39.9 μg h^{-1} mg^{-1}cat	1.22	−0.25	[54]
	BD-Ag/AF	2.07×10^{-11} mol s^{-1} cm^{-2}	7.36	−0.6	[55]
	Fe_3O_4/Ti	5.6×10^{-11} mol s^{-1} cm^{-2}	2.6	−0.4	[56]
	Mn_3O_4 @rGO	17.4 μg $h^{-1}$$mg^{-1}$cat	3.52	−0.85	[57]
Carbon based	Boron doped graphene	9.8 mg $h^{-1}$$cm^{-2}$	10.8	−0.5	[58]
	Carbon doped nitrogen	1.40 mmol g^{-1} h^{-1}	1.42	−0.9	[59]
	NP–C–MOF–5	5.77×10^{-4} μg h^{-1} mg^{-1}cat	–	−0.1	[60]
	PCN	8.09 mgh^{-1} mg^{-1}cat	11.59	−0.1	[61]
	B_4C nanosheet	26.57 μg h^{-1} mg^{-1}cat	15.95	−0.75	[62]

(continued)

Table 1 (continued)

Material	Catalyst	NH_3 yield	FE (%)	Potential (V vs. RHE)	Refs.
	Cr_2O_3–rGO	33.3 μg h^{-} mg^{-1}cat	7.33	−0.7	[63]
	PdCu/rGO	2.80 μg h^{-1} mg^{-1}cat	3.7	−0.2	[64]
	α-Au/CeO_x–rGO	8.3 μg h^{-1} mg^{-1}cat	10.1	−0.2	[65]
Single atom	Ru SAs/N–C	120.9 μg h^{-1} mg^{-1}cat	29.6	−0.2	[66]
	LiN_3	$(7.9 \pm 1.6) \times 10^{-9}$ mol cm^{-1} s^{-1}	18.5	–	[67]
	Bi nanosheet	13.23 μg h^{-1} mg^{-1}cat	10.46 ± 1.45	−0.8	[68]
Noble atom	Ru nanoparticle	21.4 mg h^{-1} m^{-2}	5.4	−100	[69]
	Rh nanosheet	23.88 μg h^{-1} mg^{-1}cat	0.217	−0.2	[32]
	Au microstructure	25.57 μg h^{-1} mg^{-1}cat	6.05	−0.2	[70]
	Ag nanosheet	4.62×10^{-11} mol s^{-1} cm^{-2}	4.8	−0.6	[71]
	RuPt alloy	5.1 10^{-9} g s^{-1} cm^{-2}	13.2	−0.123	[72]

states results in altered structure–property relationship, thereby tuning electrocatalytic efficiency [10]. The commonly explored noble metals with catalytic efficiency are Ru, Rh, Pd, Au and Ag [71]. Earlier studies have explored the electrocatalytic property of hybrid material consisting of noble metal (Pd, Ru, Rh, Pt, Au or Ag) by dispersing graphite oxide on the respective metal salts. The hybrid materials were found to enhance the heterogeneous electron transfer owing to the surface defects on the hybrid materials [73]. Ru nanoparticles have been used as efficient electrocatalyst in NRR condition with NH_3 yield rate of 21.4 mg h^{-1} m^{-2} at 60 °C. Ru NPs were anchored on a carbon fiber paper (CFP) synthesized via oleate-assisted thermal decomposition/reduction method. The N_2 adsorption was favored by the side on adsorption on two Ru atoms promoting the transfer of the first electron pair [69]. Ultrathin 2D Rh nanosheets have been synthesized via cyanogel reduction method that acts as an efficient electrocatalyst in NRR at ambient conditions. NH_3 yield of 23.88 μg h^{-1} mg^{-1}cat. has been obtained at a low potential of −0.2 V versus RHE due to its high specific surface area and modified electronic structure with numerous defective atoms [32]. Surfactant-free synthesis of an Rh flower-like structure was carried out via hydrothermal reduction of Rh precursor. The excellent catalytic activity during hydrogenation suggests that the material is an effective electrocatalyst in NRR [74]. Efficient electrocatalytic property for NRR was harbored by a flower-like Au microstructure with NH_3 yield of 25.57 μg h^{-1}

mg^{-1} at potential −0.2 V versus RHE with FE of 6.05% owing to the available surface active sites of Au atoms [70]. Molecular dynamics simulation studies show that Ru single atom anchored on carbon material graphene as an efficient electrocatalyst in NRR. The DFT calculations provide the four possible mechanisms of dissociation as depicted below:

(a) *NNH species attack by the first $H^+ + e^-$ pair results in the dissociation of NN bond;
(b) Dissociation into *N and $*NH_2$ from $*NNH_2$ species;
(c) $*NHNH_2$ decomposes to *NH and $*NH_2$;
(d) Generation of two $*NH_2$ from $*NH_2NH_2$.

During associative mechanism, the upper N atom adds electron to N_2 by forming *NNH as intermediate species [75]. Nanosheet of Ag has been explored for electrocatalytic N_2 fixation with a yield rate of NH_3, 4.62×10^{-11} mol s^{-1} cm^{-2} at voltage −0.60 V versus RHE with FE of 4.8% under ambient conditions. The material was found with high electrochemical stability and durability up to 24 h of electrocatalysis [71]. A hybrid ruthenium–platinum (RuPt) alloy was developed and found to achieve NH_3 yield rate of 5.1×10^{-9} g s^{-1} cm^{-2} with a 13.2% FE at 0.123 V versus RHE at ambient conditions. The synergistic effect of Ru and Pt mediated the high yield of ammonia with high selectivity and stability up to 48 h [72].

2.5 Transition Metal

Transition metal-based catalysts are found with d-orbital electrons which carry out π-back donation process that makes them efficient catalyst lowering the kinetic energy for activation of N≡N bond. Fe is one of the most commonly used cost effective and environmental friendly transition metal NRR catalyst. Till date, various studies have been carried out on oxides and composites of Fe that have been efficiently used as electrocatalyst for NRR. In a recent study, γ–Fe_2O_3 nanoparticle has been reported as an efficient NRR catalyst in alkaline media with highest rate of faradaic efficiency (FE) of 1.96% and NH_3 formation of 58.9 nmol h^{-1} cm^{-2} at 0.0 V RHE. Here, anion-exchange membranes electrode assembly (MEA) were used for ion-conduction with N_2-saturated basic aqueous solution that resulted in a three-fold increase in the catalytic activity of the catalyst [42]. In a different study on Fe-based nanostructure for NRR, Fe_2O_3 nanorods were found to be efficient electrocatalyst for NRR with a NH_3 yield rate of 15.9 $mgh^{-1}mg^{-1}$catalyst and FE of 0.94%. The electrocatalyst showed efficiency at a moderate temperature and atmospheric pressure operating at a voltage of 0.8 V versus RHE [43]. Molybdenum (Mo)-based electrocatalyst has been studied for NRR reaction. The catalytic ability of tetragonal Mo_2N for the NRR was investigated and synthesized via urea glass route method wherein titanium was used as a substrate for catalyst loading. The electrocatalyst was found with a FE of 42.3% obtained at −0.05 V versus RHE and NH_3 yield of 1.99 μg mg^{-1} h^{-1} [44]. Molybdenum disulfide (MoS_2) nanosheet array was found with NH_3 yield rate of 8.08×10^{-11} mol s^{-1} cm^{-1} in 0.1 m Na_2SO_4 at high FE (1.17%) and voltage at −0.5 V versus RHE. The DFT calculations indicated the role of positively charge Mo edge to polarize and activate N_2 molecules. The reductive protonation of adsorbed N_2, which is the potential-determining step (PDS), was achieved at an energy barrier of 0.68 eV which is much lower to the usual range of 1–1.5 eV [45]. Defect-rich MoS_2 nanoflower has been reported as efficient electrocatalyst with a NH_3 yield rate of 29.28 μg h^{-1} mg^{-1}cat. at voltage −0.40 V versus RHE. On the other hand, defect-free MoS_2 was found with much lower average NH_3 yield (13.41 μg h^{-1} mg^{-1}cat.) and FE (2.18%) at −0.50 V versus RHE. Based on the density functional theory (DFT) calculation, the association mechanism behind DR-MoS_2 electrocatalytic activity shows distal associative pathway to be more favorable with a free energy of 1.59 eV. During proton–electron coupling transferring process, the lower energy barrier (0.68 eV) of PDS can be related to the more positive NRR potential [46]. An efficient electrocatalyst for NRR reactions and Mn_3O_4 nanocube was synthesized via hydrothermal route with superior electrochemical stability and durability. NH_3 yield rate of 11.6 μg h^{-1} mg^{-1}cat. at voltage −0.8 V versus RHE was achieved in 0.1 m Na_2SO_4 aqueous solution at FE of 3.0%. The high electrocatalytic NRR efficacy of Mn_3O_4 can be attributed to the strong N_2 triple bond activated due to the distortion of the atomic lattice structures of Mn_3O_4 [47]. Oxides of Nb and Nb_2O_5 nanofibers have been developed as high-performance electrocatalyst with NH_3 yield of 43.6 mg$h^{-1}$$mg^{-1}$cat. at F.E of 9.26%. Here, N_2 is adsorbed on Nb_2O_5 surface in the Nb edge atom, and hydrogenation follows the distal pathway [48]. On the other hand, multishelled hollow Cr_2O_3 microspheres (MHCMs) have been used to bring about NH_3 yield of 25.3 μgh^{-1} mg^{-1}cat. wherein hydrogenation is achieved via distal association and partial alternative association pathway [49]. Electrocatalytic activity of CeO_2 nanorods reduced with hydrogen (r-CeO_2 nanorods) had been analyzed, and 16.4 μg h^{-1} mg^{-1}cat. NH_3 yield rate at −0.5 V versus RHE and FE of 3.7% at −0.4 V versus RHE had been achieved. The electrocatalytic efficiency CeO_2 owes to the Ce^{3+} and Ce^{4+} oxidation state transitions that act as catalytic site for N_2 adsorption and $*NHNH_2$–$*NH_2NH_2$ being the potential rate determining step with energy barrier ~0.77 eV. Also, the catalytic efficiency has been enhanced by oxygen vacancy defects which influence the electronic structures of metal oxides [50]. Nitrides of transition metals have been reported to show an electrocatalytic property for NRR. For instance, cubic vanadium (III) nitride and niobium (III) nitride supported on carbon cloth have been explored for electrocatalytic properties under ambient conditions [76]. Gold (Au) is studied as one of the efficient electrocatalysts for NRR and is found with low hydrogen evolution reaction efficacy. Tetrahexahedral gold nanorods (THH Au) have shown electrocatalytic NRR property with a high production yield of 1.648 μg h^{-1} cm^{-2} at −0.2 V versus RHE at ambient conditions. The catalytic mechanism starts with the chemisorption of N_2 on the Au crystal facets that act as the active sites following which activated H from H_2O electrocatalysis forms stable N–H bond. This is followed by the hydrogenation of N_2 from the electrolyte and electrode [51]. Au thin film was studied for NRR wherein the reduction mechanism involves associative pathway simultaneously with the N_2 dissociation and N atoms hydrogenation [77]. In a separate study, Au hollow nanocage (AuHNCs) was synthesized wherein the optimal size of pores and density in the walls were investigated that show the highest electrocatalytic NRR activity. The chronoamperometry (CA) tests, conducted at −0.4 V versus RHE, have shown the highest ammonia yield to be at a rate of 3.74 $\mu g\ cm^{-2}\ h^{-1}$ and FE 35.9%. For AuHNCs, it was found that the pore size of the electrocatalyst is inversely proportional to the selectivity property of NRR due to the fluctuation in surface area and the ineffective reactants captivity within the HNCs cavity [52].

In recent times, various hybrid composites have been synthesized that have been found with efficient electrocatalytic NRR efficacy. Amorphous $Bi_4V_2O_{11}$–CeO_2 hybrid was synthesized for the use as catalysts in NRR with high NH_3 yields of 23.21 $mgh^{-1}mg^{-1}$cat. and FE of 10.16%. In this case, the high electrocatalytic activity of the composites was attributed to the amorphous nature that indicates a distorted structure with dangling bonds and the significant defects that act as active sites which can reduce the energy barrier. Further, rapid interfacial charge transfer is facilitated by the introduction of band alignment by CeO_2 e with $Bi_4V_2O_{11}$ [53]. Bimodal nanoporous PdCu alloys with 3D hierarchical interconnected porous network and tailored Pd/Cu atomic ratios were studied for electrocatalytic NRR with high reduction of NH_3 from N_2 with a yield rate of 39.9 $\mu g\ h^{-1}\ mg^{-1}$cat. at ambient conditions. The structural integrity of the alloy Pd_3Cu_1 has been found to facilitate its high stability during NRR keeping it 100% efficient up to 18 h of operation [54]. Bromide-derived Ag Porous film (BD-Ag/AF) has been developed via electrochemical reduction of AgBr film on Ag foil (AgBr/AF) as a highly effective NRR electrocatalyst with a high FE of 7.36% at −0.6 V versus RHE in 0.1 N Na_2SO_4. The high efficiency owes to the fact that Ag is very easily reduced from AgBr due to its unstable nature, and the Br^- ions are facilitated to be adsorbed onto the electrode that suppresses hydrogen evolution reaction (HER). The BD-Ag/AF was found with high electrochemical stability with efficient electrocatalytic activity up to 10 consecutive cycles [55]. Ti mesh with a spinel Fe_3O_4 nanorod (Fe_3O_4/Ti) displays efficient NRR electrocatalytic property with high stability that achieved NH_3 yield of 5.6×10^{-11} mol $s^{-1}\ cm^{-2}$ under in 0.1 M Na_2SO_4, with a high FE of 2.6% and at −0.4 V versus a RHE [56]. Mn_3O_4 nanoparticles@rGO (Mn_3O_4@rGO) composite was synthesized to function as an efficient electrocatalyst for NRR that achieved 17.4 $\mu g\ h^{-1}mg^{-1}$cat. NH_3 yields at a FE of 3.52% and potential at −0.85 V versus RHE while displaying excellent selectivity and electrochemical stability. The Mn_3O_4@rGO composite was found with high transfer rate of charge resulting in rapid NRR kinetics when compared to native Mn_3O_4. Further, the binding energies calculated from DFT analysis indicate bonding between N_2 molecule and the Mn atoms [57].

2.6 *Single-Atom Catalyst*

In recent times, the use of single-atom catalyst (SACs) e for various electrochemical processes has significantly increased owing to its high specificity and catalytic activity. In SACs, the single atom acts as the active center dispersed on a support, which results in a distinct electronic structure and low coordination state with a 100% utilization of the active sites [78, 79]. One such electrocatalyst was designed with single-atom Mo (Mo-SAs) on N-doped carbon wherein the two carbon atoms and one nitrogen held the single metal atoms. The synthesized catalyst was found with two distinct kinds of active sites, Mo–N_2C_1 and Mo–N_3, both possessing the same adsorption efficiency toward N_2 and was found to enhance with the increase in the coordinated N atom number (from Mo–N_1C_2 to Mo–N_2C_1 to Mo–N_3). During hydrogenation, electrons contribution to graphene and *N_2H is carried out by Mo–N_1C_2 displaying both distal and alternative associative pathway [78]. The idea of low cost and highly efficient catalyst mediated the synthesis of Fe atom catalyst embedded on Nx-doped ($x = 1, 2, 3$) graphene wherein adsorption of N_2 molecule took place in two different ways via N_2 positioned parallel to the graphene plane on top of Fe atom and vertically against graphene plane through the single atom N. The energy of N atom adsorption is strengthened by the Fe magnetic moment that mediates the transfer of charge between N_2 molecule and substrate that results in the lower potential of NRR [80]. Geng with his group in 2018 developed nitrogen-doped carbon with single-atom Ru (Ru SAs/N–C) by pyrolyzing zeolitic imidazolate which showed a high catalytic performance of NH_3 yield of 120.9 $\mu g\ h^{-1}\ mg^{-1}$cat. with FE 29.6% at −0.2 V versus RHE. The mechanism of NRR followed the distal associative pathway, and the Ru atoms acted as the active sites for N_2 adsorption with a turnover frequency (TOF) number 376 h^{-1} based on surface Ru atoms. The electrochemical stability of Ru SAs/N–C was found to be static up to 12 h, and the Ru single atoms were dispersed after electrolysis [66]. A systematic investigation based on DFT calculation was carried out for transition metal atoms (Sc to Zn, Mo, Ru, Rh, Pd and Ag) anchored on defective boron nitride (TM–BN) as an analog to graphene for NNR activity. Boron nitride monolayer with Mo atom with the boron monovacancy exhibited better performance in adsorption of N_2 and destabilization of NH_2 species that results in the reduction of overpotential of PDS of 0.19 V [81]. In a similar study, graphitic carbon nitride-based SACs anchoring with five single atoms (Ti, Co, Mo, W and Pt) have been investigated using DFT computations for electrochemical nitrogen reduction. The SACs with highest electrocatalytic activity were assessed by analyzing the free energy changes and PDS for different pathways. The adsorption of N_2 on metal atom followed end-on configuration for all the SACs, and W@C_3N_4 showed the highest catalytic activity with the lowest negative potential of −0.35 V and associative enzymatic pathway [15]. Models have been constructed using single boron atom attached to 2-D materials, viz. boron sulfide (BS), boron nitride (BN), graphene (G), S-triazine-based g-C_3N_4 (S), black phosphorus (BP), h-MoS_2 (h), tri-s-triazine-based g-C_3N_4 (T) and T-MoS_2 (t) to investigate for their elctrocatalytic property for N_2 reduction. The models have been designed for three boron states—substituted (S), lattice boron (D) and adsorbed (A). The calculated ΔG values for energy barriers indicate catalysts hS1 and G-A as efficient in NRR with small activation barriers of 0.31 and 0.46 eV. The DFT calculations showed the distal associative pathway, favored in the NRR mechanism [82]. β12 phase of the boron monolayer (BM) anchored with V atom (V/β12-BM) was investigated for electrocatalytic property based on MD simulation. The composite was found with a very low overpotential of 0.28 V. This low potential owes to the fact that during side on N_2 adsorption on V/β12-BM, N_2 1π orbital gets ionized elongating the N–N bond and decreasing its strength. The designed SAC was also found with high thermodynamic stability and was effective in suppressing HER during NRR (Z [83]. Studies have been reported on lithium-mediated electrocatalytic N_2 reduction, spontaneously splitting nitrogen triple bond to form lithium nitride (LiN_3). The highest NH_3 yield rate was found to be $(7.9 \pm 1.6) \times 10^{-9}$ mol cm^{-1} s^{-1} at FE 18.5% [67]. A 2D-mosaic bismuth nanosheet (Bi–NS) was synthesized for electrochemical reduction of N_2 that resulted in NH_3 yield of ~13.23 μg $h^{-1}mg^{-1}$cat. with FE 10.46 ± 1.45% at −0.8 V versus RHE. The p-orbital electron delocalization of bismuth and its semi-conductive property limiting the accessibility of surface electron results in the high FE indicating that structure modulation greatly affects the electrocatalytic property of synthesized catalyst [68]. Ru single atom sites anchored in N-doped porous carbon were synthesized using a coordination-assisted strategy that enabled high NH3 yield rate of 3.665 at −0.21 V versus RHE. The coordination of ligands was tuned that enabled high catalytic efficiency [84].

2.7 Carbon-Based Composites

In recent times, carbon-based materials have been explored for their electrocatalytic properties owing to their cost effective, environment-friendly and noncorrosive nature and are electrochemically stable with a high surface area. Also, carbon materials are found with high conductive nature with hierarchical pore structure facilitating electron transfer [69, 85]. Tuning of catalytic activity of these materials is achieved by doping with heteroatoms, thereby creating more surface defects to modulate energy band gap, spin density and charge density [86, 87]. Boron-doped graphene (BG) is one such materials synthesized by thermally annealing graphene oxide and boric acid for electrochemical NRR. The dopant boron have been found to enhance electrocatalytic property of graphene with NH_3 yield of 9.8 mg $h^{-1}cm^{-2}$ and the FE 10.8% at −0.5 V versus RHE. The electron-deficient environment at the position of boron doping facilitates the binding of N_2, and the BC_3-type bond structure has been found with highest catalytic activity due to the lowest reaction energy barrier of 0.43 eV [58]. Porous carbon doped with nitrogen (NPC) has been used for electrochemical NRR. The pyridinic and pyrrolic N act as primary active sites for the N_2 adsorption and dissociation. The hydrogenation pathway calculated from DFT analysis was found to be $*N{\equiv}N \rightarrow *NH{\equiv}NH \rightarrow *NH_2{-}NH_2 \rightarrow 2NH_3$, and the rate of NH_3 was 1.40 mmol g^{-1} h^{-1} at potential −0.9 V versus RHE [59]. Carbon-based catalyst co-doped with N and P (NP–C–MOF–5) has been found with electrocatalytic activity for NRR under atmospheric pressure and temperature. NH_3 yield of 5.77×10^{-4} μg h^{-1} mg^{-1}cat. was achieved at −0.1 V versus RHE. The N_2Hy formation and detection of $N_2H_4 \cdot H_2O$ in the FTIR studies suggest the associative pathway as the probable mechanism of NRR by NP–C–MOF–5 [60]. In the context of defect engineering to develop electrocatalyst, polymeric carbon nitride (PCN) matrix is synthesized to introduce nitrogen vacancy defects which modulate π electron delocalization that redistributes the extra electrons to adjacent carbon atoms, thereby enhancing N_2. The achieved NH_3 yield was 8.09 $mgh^{-1}mg^{-1}$cat. and FE of 11.59% [61]. Sulfur dots–graphene nanohybrid has been found with appreciable electrocatalytic activity with NH_3 yield of 28.56 μg h^{-1} mg^{-1}cat. and FE 7.07% at −0.85 V versus RHE in $LIClO_4$. The electrocatalyst was found with high electrochemical stability up to 24 h of electrolysis [88].

Boron carbide (B_4C) nanosheets have been found with electrocatalytic efficiency under atmospheric pressure and temperature in 0.1 M HCL. The achieved NH_3 formation rate and FE are 26.57 μg h^{-1} mg^{-1}cat. and 15.95%, respectively, at −0.75 V. This aqueous-based NRR electrocatalysts performed catalytic activity with stability up to 30 h. For both the end-on and side-on configuration of N_2 adsorbtion, wherein one and two terminal N atom is bonded to the B atom on the B_4C surface, the adsorption potential energies were calculated to be almost similar 0.65 eV and 0.63 eV, respectively [62]. Evenly dispersed Cr_2O_3 with rGO has been developed that was found to achieve electrocatalytic activity in NRR with highly efficient selectivity and electrochemical stability. The reaction mechanism follows the distal association pathway and the even dispersion on rGO enhances the adsorption sites on the hybrid. The electrocatalyst resulted in a large NH_3 yield of 33.3 μg h^{-1} mg^{-1}cat. and FE of 7.33% at −0.7 and −0.6 V versus RHE [63]. In a study on rGO-based bimetallic alloy, PdCu/rGO composite had been synthesized by anchoring PdCu amorphous nanoclusters on rGO, and NH_3 yield rate of 2.80 μg h^{-1} mg^{-1}cat. has been achieved. The synthesis was carried out by coreduction of GO and the metal precursors under controlled size and composition. The catalytic efficiency of the hybrid composites wase found to be influenced by the stoichiometric ratio of Pd or Cu precursors owing to the active sites available on Pd and Cu. However, the addition of rGO as supporting matrix greatly enhances N_2 adsorption, thereby modulating NRR activity by the dispersion of catalytic active sites [64]. A α–Au/CeOx–RGO hybrid nanocomposite has been developed following coreduction route, and ammonia yield rate of 8.3 μg h^{-1} mg^{-1}cat. is achieved with a FE of 10.10% and at −0.2 V versus RHE. Here, the presence CeOx mediates amophization of Au nanoparticles owing to the strain around Ce^{3+} and Ce^{4+} that enhance the electrocatalytic activity as compared to crystalline Au due to structural distortion [65].

3 Conclusion

Electrochemical nitrogen fixation has achieved greater heights owing to the development of various heterogeneous electrocatalysts with commendable efficiency of nitrogen-to-ammonia conversion. The developed electrocatalyst has been found to possess catalytic property with desirable characteristics of cost effectivity and eco-friendly nature. The most significant characteristic of electrocatalytic nitrogen reduction stems from the fact that electrochemical NRR is driven by renewable energy source under ambient conditions. However, the impending search for electrocatalyst with high efficiency, selectivity and durability to achieve high yield of ammonia has been on process wherein design engineering of the materials have been extensively carried out to develop new electrocatalysts. As discussed, MD simulation-based DFT calculation provides valuable insights into the screening and designing of reliable electrocatalyst for NRR and enables better understanding of the catalytic mechanism and the thermodynamics of NRR. To summarize, electrochemical nitrogen reduction provides an efficient means to produce NH_3 which can be regarded as an alternate means to Haber–Bosch process.

References

1. Suryanto BH, Wang D, Azofra LM, Harb M, Cavallo L, Jalili R, MacFarlane DR (2018) ACS Energy Lett 4(2):430–435
2. Cao N, Zheng G (2018) Nano Res 11(6):2992–3008
3. Hoffman BM, Lukoyanov D, Yang ZY, Dean DR, Seefeldt LC (2014) Chem Rev 114(8):4041–4062
4. Van der Ham CJ, Koper MT, Hetterscheid DG (2014) Chem Soc Rev 43(15):5183–5191
5. Guo X, Du H, Qu F, Li J (2019) J Mater Chem A 7(8):3531–3543
6. Mukherjee S, Cullen DA, Karakalos S, Liu K, Zhang H, Zhao S, Wu G (2018) Nano Energy 48:217–226
7. Deng J, Iñiguez JA, Liu C (2018) Joule 2(5):846–856
8. Lindley BM, Appel AM, Krogh-Jespersen K, Mayer JM, Miller AJ (2016) ACS Energy Lett 1(4):698–704
9. Guo C, Ran J, Vasileff A, Qiao SZ (2018) Energy Environ Sci 11(1):45–56
10. Wan Y, Xu J, Lv R (2019) Mater Today. (In press)
11. Seh ZW, Kibsgaard J, Dickens CF, Chorkendorff IB, Nørskov JK, Jaramillo TF (2017) Science 355(6321):eaad4998
12. Peterson AA, Nørskov JK (2012) J Phys Chem Lett 3(2):251–258
13. Doyle AD, Montoya JH, Vojvodic A (2015) ChemCatChem 7(5):738–742
14. Halck NB, Petrykin V, Krtil P, Rossmeisl J (2014) Phys Chem Chem Phys 16(27):13682–13688
15. Chen Y, Ji S, Chen C, Peng Q, Wang D, Li Y (2018) Joule 2(7):1242–1264
16. Mistry H, Varela AS, Kuehl S, Strasser P, Cuenya BR (2016) Nat Rev Mater 1(4):16009
17. Qian J, Shen M, Zhou S, Lee CT, Zhao M, Lyu Z, Xia Y (2018) Mater Today 21(8):834–844
18. Xiao W, Lei W, Gong M, Xin HL, Wang D (2018) ACS Catal 8(4):3237–3256
19. Nazemi M, El-Sayed MA (2018) J Phys Chem Lett 9(17):5160–5166
20. Yang Y, Luo M, Zhang W, Sun Y, Chen X, Guo S (2018) Chem 4(9):2054–2083
21. Wang X, Gan X, Hu T, Fujisawa K, Lei Y, Lin Z, Lv R (2017) Adv Mater 29(4):1603617
22. Liu HM, Han SH, Zhao Y, Zhu YY, Tian XL, Zeng JH, Chen Y (2018b) J Mater Chem A 6(7):3211–3217
23. Zhou Z, Li Y, Xia M, Zhou N (2018) Solid State Ionics 315:1–6
24. Li H, Shang J, Ai Z, Zhang L (2015) J Am Chem Soc 137(19):6393–6399
25. Wendt S, Sprunger PT, Lira E, Madsen GKH, Li Z, Hansen JO, Matthiesen J (2008) Science 320:1755–1759
26. Zhang N, Li X, Ye H, Chen S, Ju H, Liu D, Zhu J (2016) J Am Chem Soc 138(28):8928–8935
27. Deng D, Chen X, Yu L, Wu X, Liu Q, Liu Y, Si R (2015) Sci Adv 1(11):e1500462
28. Zhang N, Jalil A, Wu D, Chen S, Liu Y, Gao C, Wu X (2018b) J Am Chem Soc 140(30):9434–9443
29. Jin H, Guo C, Liu X, Liu J, Vasileff A, Jiao Y, Qiao SZ (2018) Chem Rev 118(13):6337–6408

30. Peng Y, Lu B, Chen S (2018) Adv Mater 30(48):1801995
31. Shen M, Wei C, Ai K, Lu L (2017) Nano Res 10(5):1449–1470
32. Liu G, Robertson AW, Li MMJ, Kuo WC, Darby MT, Muhieddine MH, Tsang SCE (2017) Nature Chem 9(8):810
33. Yang XF, Wang A, Qiao B, Li J, Liu J, Zhang T (2013) Acc Chem Res 46(8):1740–1748
34. Zhao C, Dai X, Yao T, Chen W, Wang X, Wang J, Li Y (2017a) J Am Chem Soc 139(24):8078–8081
35. Bai S, Wang L, Li Z, Xiong Y (2017) Adv Sci 4(1):1600216
36. Jao MH, Lu CF, Tai PY, Su WF (2017) Crys Growth Des 17(11):5945–5952
37. Sun S, Zhang X, Yang Q, Liang S, Zhang X, Yang Z (2018) Prog Mater Sci 96:111–173
38. Hong JW, Kim Y, Kwon Y, Han SW (2016) Chem–An Asian J 11(16):2224–2239
39. Shao M, Chang Q, Dodelet JP, Chenitz R (2016) Chem Rev 116(6):3594–3657
40. Montoya JH, Tsai C, Vojvodic A, Nørskov JK (2015) Chemsuschem 8(13):2180–2186
41. Yang D, Chen T, Wang Z (2017) J Mater Chem A 5(36):18967–18971
42. Kong J, Lim A, Yoon C, Jang JH, Ham HC, Han J, Park HS (2017) ACS Sustain Chem Eng 5(11):10986–10995
43. Xiang X, Wang Z, Shi X, Fan M, Sun X (2018) ChemCatChem 10(20):4530–4535
44. Hu B, Hu M, Seefeldt LC, Liu TL (2019) ACS Energy Lett
45. Zhang L, Ji X, Ren X, Ma Y, Shi X, Tian Z, Sun X (2018a) Adv Mater 30(28):1800191
46. Li X, Li T, Ma Y, Wei Q, Qiu W, Guo H, Tang B (2018) Adv Energy Mater 8(30):1801357
47. Wu X, Xia L, Wang Y, Lu W, Liu Q, Shi X, Sun X (2018) Small 14(48):1803111
48. Han J, Liu Z, Ma Y, Cui G, Xie F, Wang F, Sun X (2018) Nano Energy 52:264–270
49. Zhang Y, Qiu W, Ma Y, Luo Y, Tian Z, Cui G, Sun X (2018c) ACS Catal 8(9):8540–8544
50. Xu B, Xia L, Zhou F, Zhao R, Chen H, Wang T, Gong F (2019) ACS Sustain Chem Eng 7(3):2889–2893
51. Bao D, Zhang Q, Meng FL, Zhong HX, Shi MM, Zhang Y, Zhang XB (2017) Adv Mater 29(3):1604799
52. Nazemi M, Panikkanvalappil SR, El-Sayed MA (2018) Nano Energy 49:316–323
53. Lv C, Yan C, Chen G, Ding Y, Sun J, Zhou Y, Yu G (2018) Angewandte Chemie Int Edition 57(21):6073–6076
54. Pang F, Wang Z, Zhang K, He J, Zhang W, Guo C, Ding Y (2019) Nano Energy 58:834–841
55. Ji L, Shi X, Asiri AM, Zheng B, Sun X (2018) Inorg Chem 57(23):14692–14697
56. Liu B, Wang Y, Peng HQ, Yang R, Jiang Z, Zhou X, Zhang W (2018a) Adv Mater 30(36):1803144
57. Huang H, Gong F, Wang Y, Wang H, Wu X, Lu W, Li T (2019) Nano Res 1–6
58. Yu X, Han P, Wei Z, Huang L, Gu Z, Peng S, Zheng G (2018) Joule 2(8):1610–1622
59. Liu Y, Su Y, Quan X, Fan X, Chen S, Yu H, Zhao J (2018c) ACS Catal 8(2):186–1191
60. Song P, Kang L, Wang H, Guo R, Wang RM (2019) ACS Appl Mater Interfaces
61. Lv C, Qian Y, Yan C, Ding Y, Liu Y, Chen G, Yu G (2018) Angewandte Chemie Int Edition 57(32):10246–10250
62. Qiu W, Xie XY, Qiu J, Fang WH, Liang R, Ren X, Tang B (2018) Nature Commun 9(1):3485
63. Xia L, Li B, Zhang Y, Zhang R, Ji L, Chen H, Liu Q (2019a) Inorg Chem 58(4):2257–2260
64. Shi MM, Bao D, Li SJ, Wulan BR, Yan JM, Jiang Q (2018) Adv Energy Mater 8(21):1800124
65. Li SJ, Bao D, Shi MM, Wulan BR, Yan JM, Jiang Q (2017) Adv Mater 29(33):1700001
66. Geng Z, Liu Y, Kong X, Li P, Li K, Liu Z, Zeng J (2018) Adv Mater 30(40):1803498
67. Lazouski N, Schiffer ZJ, Williams K, Manthiram K (2019) Joule 3:1127–1139
68. Li L, Tang C, Xia B, Jin H, Zheng Y, Qiao SZ (2019) ACS Catal
69. Wang D, Azofra LM, Harb M, Cavallo L, Zhang X, Suryanto BH, MacFarlane DR (2018) Chemsuschem 11(19):3416–3422
70. Wang Z, Li Y, Yu H, Xu Y, Xue H, Li X, Wang L (2018b) ChemSusChem 11(19):3480–3485
71. Huang H, Xia L, Shi X, Asiri AM, Sun X (2018) Chem Commun 54(81):11427–11430
72. Manjunatha R, Schechter A (2018) Electrochem Commun 90:96–100
73. Giovanni M, Poh HL, Ambrosi A, Zhao G, Sofer Z, Šaněk F, Pumera M (2012) Nanoscale 4(16):5002–5008

74. Jiang Y, Su J, Yang Y, Jia Y, Chen Q, Xie Z, Zheng L (2016) Nano Res 9(3):849–856
75. Cao Y, Gao Y, Zhou H, Chen X, Hu H, Deng S, Wang J (2018b) Advan Theory Simul 1(5):1800018
76. Du HL, Gengenbach T, Hodgetts R, Macfarlane DR, Simonov AN (2019) ACS Sustain Chem Eng 7:6839–6850
77. Yao Y, Zhu S, Wang H, Li H, Shao M (2018) J Am Chem Soc 140(4):1496–1501
78. Ling C, Bai X, Ouyang Y, Du A, Wang J (2018) J Phys Chem C 122(29):16842–16847
79. Jia M, Fan Q, Liu S, Qiu J, Sun Z (2018) Curr Opin Green Sustain Chem 16:1–6
80. Guo X, Huang S (2018) Electrochim Acta 284:392–399
81. Zhao J, Chen Z (2017) J Am Chem Soc 139(36):12480–12487
82. Liu C, Li Q, Wu C, Zhang J, Jin Y, MacFarlane DR, Sun C (2019) J Am Chem Soc
83. Zhu H, Hu YL, Wei SH, Hua D (2019) J Phys Chem C
84. Tao H, Choi C, Ding LX, Jiang Z, Han Z, Jia M, Hong S (2019) Chem 5(1):204–214
85. Zhao S, Lu X, Wang L, Gale J, Amal R (2019) Adv Mater 1805367
86. Kumar P, Vanhala TI, Törmä P (2017) Phys Rev B 96(24):245127
87. Tahir M, Pan L, Idrees F, Zhang X, Wang L, Zou JJ, Wang ZL (2017) Nano Energy 37:136–157
88. Chen H, Zhu X, Huang H, Wang H, Wang T, Zhao R, Sun X (2019) Chem Commun 55(21):3152–3155

Low-Pressure Ammonia Production

Komal P. Mehta, Rama Rao Karri and N. M. Mubarak

Abstract The major challenges encountered in ammonia production are the surface adsorbates and catalyst structures under conditions significant for ammonia synthesis. The discovery of scalable, active, and long-lived catalysts plays an important role in producing sustainable ammonia synthesis. Other factors that influence the ammonia production are the development of low-pressure and -temperature processes and the development of photochemical and electrochemical routes for N_2 reduction based on homogeneous (molecular) and heterogeneous catalysis. In this chapter, the favouring conditions for low-pressure ammonia production are reviewed and discussed in detail. Approaches to be adapted to achieve the improved yield, increase N_2/H_2 conversion to ammonia, lowering the capital costs, and consumption of less energy were discussed.

Keywords Ammonia · Catalysts · Haber–Bosch process · Quasi-compounds · Electrocatalysis and photocatalysis · Adsorption

1 Introduction

Ammonia (NH_3) is the highest contributing factor in chemical industries producing more than 140 million tons of nitrogen. The main usage of NH_3 is in fertilizer industry (>75%), besides it is also extensively used as a refrigerant gas, production of textiles, plastics, pesticides, and other significant chemicals. It is also used as a cleaning agent in household and industries. Even though ammonia production exists for many years, there are problems associated with its production in terms of

K. P. Mehta
Civil Engineering Department, ITM Universe, Vadodara, Gujarat, India

R. R. Karri (✉)
Department of Petrochemical Engineering, Faculty of Engineering, Universiti Teknologi Brunei, Bandar Seri Begawan, Brunei Darussalam
e-mail: kramarao.iitd@gmail.com

N. M. Mubarak
Department of Chemical Engineering, Faculty of Engineering and Science, Curtin University, 98009 Miri, Sarawak, Malaysia

Inamuddin et al. (eds.), *Sustainable Ammonia Production*,
Green Energy and Technology, https://doi.org/10.1007/978-3-030-35106-9_7

manufacturing and instrumentation issues. Due to that, it poses a severe threat to mankind and the environment [1].

Ammonia is the main component in the chemical industry. Known and accepted innovation of the twentieth century is the production of NH_3 by Haber–Bosch process [2]. More than 50% requirement of nitrogen for human use is met by ammonia production processes. Industrial production of NH_3-based fertilizers is met by this process and is operated at high temperature and pressure. The information represented in Fig. 1 demonstrates how the production of NH_3 correlates with the population growth during the period 1900–2015. This clearly indicates that, the NH_3 generation is increased many folds to meet the demand.

An acceptable solution as a substitute of carbon is ammonia production based on a biological one. It is not possible that the existing process can be replaced by more energy efficient biological process, but renewable sources like wind or solar may give energy. If on-demand fertilizer production near the farm can be practiced, it will help to solve problems of runoffs, nitrogen pollution related to the fertilizer industry (Fig. 2) [3].

The capital expenditure, energy consumption, and environmental impact of this well-known process have encouraged the researchers to develop a renewable and affordable process under mild conditions [4]. At high temperature and pressure, ammonia synthesis is feasible for market and also energy producing. High pressure and fast kinetics inhibit the production of the reverse reaction. Ammonia is a synthetic nitrogen fertilizer [5]. Without ammonia (fertilizer) which is hydrogen carrier and carbon-neutral liquid fuel, the availability of food for 50% population would have been a great challenge. The necessity of new catalyst is highly required as now manufacturing process for ammonia is known though high temperature and pressure are still challenge for chemical industry to solve [6, 7].

Plant life is supported mainly by ammonia, which is used for the production of protein and is also used as cleaners (glass and window, toilet bowl, waxes, oven, and drain—mainly diluted ammonium hydroxide) and in preparation of environmental

Fig. 1 Comparison of annual NH_3 production rate versus population growth. [*Data* from U.S. geological survey 2015]

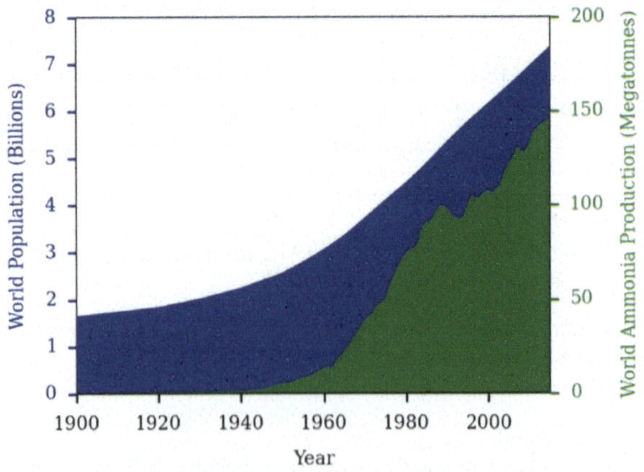

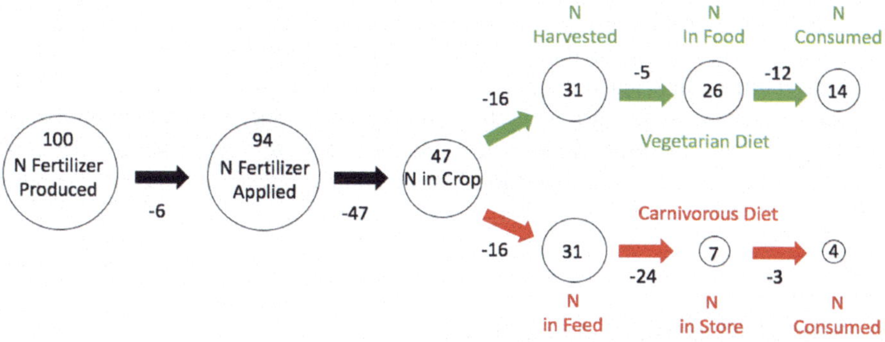

Fig. 2 Schematic diagram showing the losses and overall efficacy of nitrogen produced, delivered and uptake by carnivorous and vegetarian diets [3]

impacting things like plastic, dyes, drugs, explosives, HNO_3, and NH_4OH. Aquatic animals generate ammonia, available in dissolved form in water bodies at or near the waste site. At hazardous waste sites (attached with soil particles), ammonia gets converted into vapor without any residuals. The nitrogen cycle is balanced from animals and plants decomposition, and by excreta of animals, ammonia is returned to the aquatic system. It is one of the important pollutants to be taken care of as a toxic, lower rate of growth, or reproduction. The production of renewable NH_3 is two to three times costly than usual [8].

In this chapter, the issues faced in the production of ammonia are thoroughly discussed as well as difficulties faced in various stages of production. The Haber–Bosch process is a widely accepted method but the limitations/difficulties in this process are reviewed. The various possible alternatives for ammonia synthesis and challenges in these processes are discussed.

2 The Haber–Bosch Process

Till the twentieth century, this process was mainly used for development of production [9]. At temperature (400 °C) and pressure (>100 bars), the pure N_2 reacts with H_2 over iron catalyst takes place.

$$0.5N_2(g) + 1.5H_2(g) \leftrightarrow NH_3(g); \Delta H^\circ = -46\,kJ/mol$$

The reaction is thermodynamically favored at lower reaction temperatures. This methods develops a more active catalyst allowing a reduction in stringent conditions. Production of NH_3 by this process meets 1 to 2% of total energy demand world while [10, 11]. Economically beneficial and environmentally friendly methodology is much awaited. If more active catalyst that be operated at low pressure and temperature, with necessary kinetics which can produce equilibrium yield. It is considered as a great challenge to find a method for lower temperature and pressure. Whereas, N_2 reduction to NH_3 is performed at high temperature and pressure [12].

Catalysts that can replace the current one in Haber–Bosch process must be able to perform well under lower temperature and pressure condition. Ruthenium based catalyst have been identified to be a good catalyst [13–16]. Increasing temperature and decreasing pressure improved the hydrocarbons conversion and H_2/CO ratio [17].

3 Manufacturing of Ammonia

Ammonia is produced on a larger scale with steam reforming on natural gas. Till date, major reasons reported are reforming, and synthesis loop. Failures and technical issues do occur frequently in the NH_3 plant even after following all the standards of safety and risk assessment, which largely impacts the environment and thus market [1].

3.1 Ammonia Production

Production of ammonia is in practice by natural gas or naphtha. The four stages of ammonia production are shown in Fig. 3.

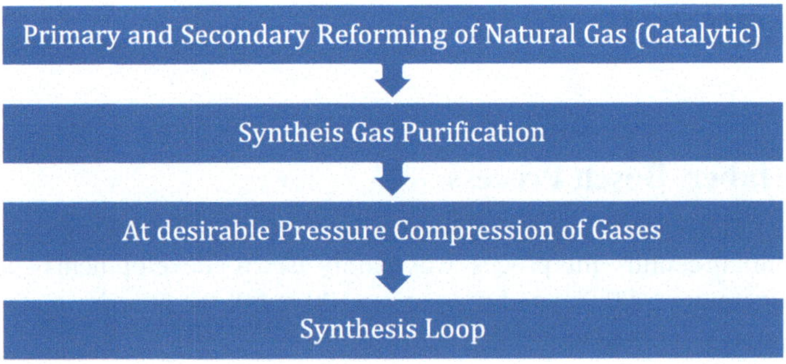

Fig. 3 Four stages of ammonia production

3.2 *Stage 1: Catalytic Reforming*

3.2.1 Desulphurization and Primary Reforming

- Steam reforming produces synthesis gas.
- Removal of sulfurous compounds with an increase in temperature at 400 °C.
- Reaction: $ZnO + H_2S \rightarrow ZnS + H_2O$.
- Passing natural gas to the primary reformer.
- Fed the steam with very high temperature into reformer with methane with required temperature tubes containing catalyst nickel.
- Tubes are heated externally due to the combustion of fuel to approximately 770 °C. CO/CO_2 is formed by CH_4 in the presence of catalysts H. Converted gas is synthesis gas. The range of operating pressure is more than 25 bars.
- Process gas mixture is to be sent to the auto-thermal reformer. Outlet of reformer contains methane 6 mol%; carbon oxides 8%; carbon dioxides 6%; H_2 50%; and water molecules 30%.

Dai [18] studied the prediction for numbers of years of usage of the material for tubes (which are under pressure) used in the manufacturing of ammonia at very high temperature greater than 800 °C. With heat treatment, the life of such tubes increased to more than double the life expected. But, due to the increase in temperature, reformer furnace can explode, which is also a failure of stress corrosion and welding [19]. Ray et al. [20] investigated failure study of tubes due to high temperature. Swaminathan et al. [21] concluded that due to changes in the design of pigtail, the end of the pipe in ammonia production plant has led to shortening of pipe and added stress to it. So the possibility of expansion of tubes decreased. It led to failure of tubes in ammonia plant over the years resulting incidents like fire [22].

3.2.2 Secondary Reformer

For entering into the secondary reformer, the produced gas is cooled up to 750 °C and gets mix with air for supplying the required quantity of N_2 for the process of synthesis gas. The reaction between methane and oxygen is highly exothermi, which favors the production of more hydrogen. The main chemical reactions that take place in the process are:

$$2O_2 + CH_4 \rightarrow 2H_2O + CO_2$$

$$O_2 + CH_4 \rightarrow CO_2 + 2H_2$$

$$O_2 + 2CH_4 \rightarrow 2CO_2 + 4H_2$$

$CO_2 + H_2O \rightarrow CO_2 + H_2$

Heat is required in the secondary reformer in the air mixing process. Water, CO_2, CO, CH_4, and catalysts are required to be removed from the gas stream. Iron oxidation can be prevented by the removal of gas from the stream which may occur in shift conversion, CO_2 removal, and methanation. Metal dusting possibility will be high due to the increase in the rate of carburization. Temperature and gas composition falls in the range of carburization; metallic components may be degraded in parts of reforming [23].

In newly developed techniques, the fire tube furnace can be replaced by heat exchanger reformer, so the integration of energy is of a good level [24]. Singh et al. [25] explained the study of fire in the plant of ammonia production. The pressure shell of interconnecting pipe of reformers was the responsible factor for the same. The accident stopped the plant for 15 days, and a detailed study was done for ruptured pipe and failure. Due to short-term high-temperature stress rupture, local damage took place, which led to failure. The possibility of ignorance of voids and cracks was also explained. Repair work in boiler and nozzle area gave a shock of failure of refractory. Jahromi et al. [26] presented a case study of downstream of secondary reformer failure of old waste boiler tubes.

3.3 Stage 2: Synthesis Gas Purification

3.3.1 Shift Conversion

In stage 2, the reformer will be followed by water gas shift reactors for converting CO to CO_2, which can be used in the synthesis of urea. To get comfortable equilibrium in water gas shift, iron oxide at 400 °C is followed by 200 °C on a copper-based catalyst.

3.3.2 CO_2 Removal

Purification of synthesis gas is taken care by removal of CO_2 (reduction to 5–10 ppm) by absorption. Absorption is with hot K_2CO_3, Selexol (most commonly used), or MEA [27]. Simple flashing can regenerate with lower consumption of energy. Pure carbon dioxide is sent to urea plant for compression after removing water. Carbon steel walls are corroded by CO_2, which also has a solution for processes. The major reason of corrosion is given in Fig. 4.

Harjac et al. [28] explained the main reasons that caused accidents/incidents of corrosion in hot K_2CO_3 acid gas removal plant. Pande et al. [29] presented a case study that hydrogen presence can cause destruction of pipeline, while the transfer of pipeline for CO_2 from ammonia plant to urea plane.

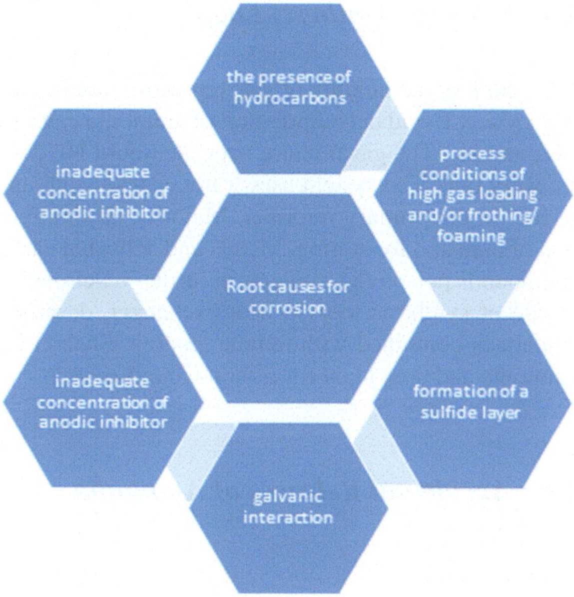

Fig. 4 Reasons for corrosion in ammonia production process

3.3.3 Methanation

Pretreatment of removal of carbon gases is performed using methanator. This is purifying process before it enters to ammonia converter. In the design of tubes, factors to be considered are balancing heat transfer, optimization of catalyst volume, and minimization of thermal volume. In a conventional plant of ammonia production, these factors are major issues. Temperature increase or decrease may result in more leakage of carbon monoxide from shift converter, which passes from a unit of absorption, thus increases the temperature in methanator [30].

3.4 Stage 3: Compression Process

After the removal of condensed water, the mixture of gas is cooled and compressed in the centrifugal compressor at a pressure of 150–300 bar with decrease in temperature. They are operated by steam turbines with the use of self-generated steam, which is also significant approach for reduction of energy usage in the plant. Krivshich et al. [31] discussed the possibility of maximizing the reliability of compressors with dry gas seal (DGS) systems.

3.5 Stage 4: Synthesis Loop

After cooling the mixture of gas, removal of ammonia (liquid form) from separator takes place. Rapid decompression of ammonia causes production of byproducts like CH_4 and H_2. The gas mixture over the liquid NH_3 is removed and sent back to the recovery unit. The system uses water as a solvent, and the gases remained are used as fuel in the primary reformer. Majority of ammonia is mostly used for production of urea and for storing. Metals and activated carbon exhibit high efficiency, and compounds poison the catalysts in further reactions. Anwar et al. [32] discussed the failure of synthesis loop in heat exchanger, describing the possibilities of hydrogen attack combined with fatigue loading. Shah et al. [33] gave useful designs for ammonia and urea plant for ribbon wound vessels.

3.6 Accidental Releases of Ammonia

Toxicity of Ammonia

Ammonia has toxicity, used on a larger scale; lower molecular weight makes it possible to form denser, so ammonia can control the hazard as it has such an important property to control it [34].

Dispersion into atmosphere

Preferable storage of NH_3 is in iron or steel container as it can be corrosive to metals. There is possibility of flash fire, if it is stored as a liquefied gas. Depending on temperature conditions, it can be less dense and forms clouds.

Other Important Studies

Khan et al. [35] presented a case study of the analysis of the failure of ammonia by fitting Weibull distributions and simulation modeling technique. With the use of existing plant configurations, they estimated the availability of plant and improvement possibility by modification or change of strategy.

4 Electron/Proton Driven Processes for NH_3 Production

4.1 Photocatalysis and Electrocatalysis

Figure 5 depicts the production of NH_3 with N_2RR catalysts (a direct electrocatalytic or photocatalytic) by oxidizing H_2O to release electrons and protons for the N_2RR in production of ammonia.

The overall reaction is $2N_2 + 6H_2O \rightarrow 4NH_3 + 3O_2$.

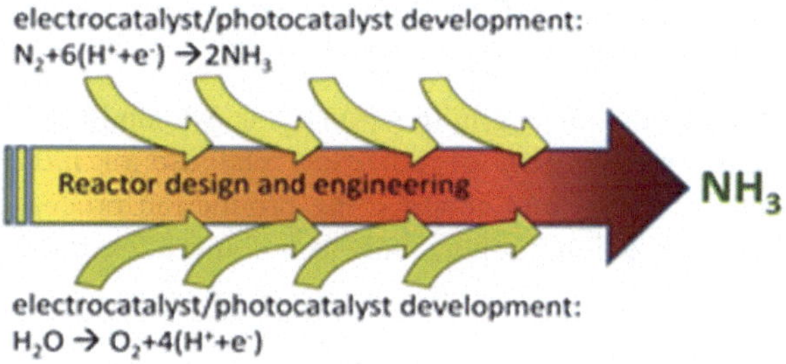

Fig. 5 Direct electrocatalytic/photocatalytic pathway to NH3 production

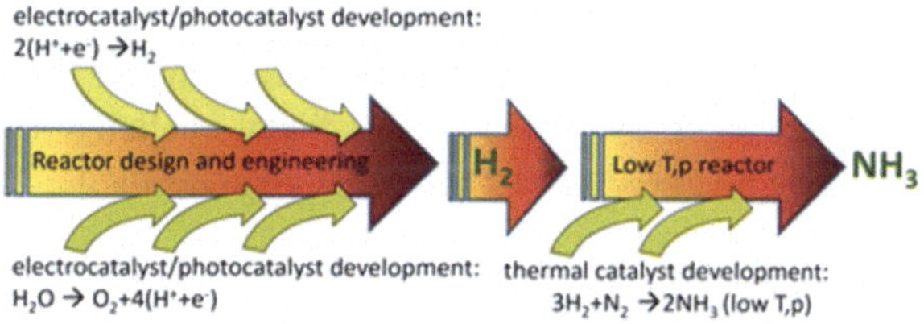

Fig. 6 Indirect electrocatalytic/photocatalytic pathway for NH3 production

A thermally catalyzed NH_3 production process is coupled with electrocatalysis or photocatalysis to produce renewable H_2 (See Fig. 6). Here, the reactions taking place are

$$2H_2O \rightarrow 2H_2 + O_2$$

$$N_2 + 3H_2 \rightarrow 2NH_3$$

In existing conventional process with the replacement of H_2, which is normally derived by renewable sources. It can reduce emissions of CO_2 [36–39].

4.2 *Electrochemical NH_3 Synthesis*

With less efficiency, N_2 is reduced to NH_3. It is a challenge to stop contamination of ammonia with other sources. Till date, the literature supports ammonia production with contamination. To resolve the stated problem, ammonia is synthesized in the presence of argon.

4.3 Photocatalytic NH_3 Synthesis

The major challenge in this process is availability of high-energy UV light requirement for reduction of nitrogen. The use of more efficient catalysts can subside the issues and enhance the production process. For a photo-electrochemical characterization, proper catalysts and more feasibility studies are required.

4.4 Electrocatalytic Oxygen Evolution Reaction

To meet the deficiency of kinetic losses, higher potentials are required in spite of catalytic materials. This is a significant challenge to overcome, and new catalysts that are able to reduce the scaling are yet awaited.

4.5 Electrocatalytic Hydrogen Evolution Reaction

Compared to NH_3, one hydrogen evolution reaction (HER) method is also practiced for aqueous electrochemical systems. It controls the catalytic surface by promoting the reduction of dinitrogen. Mitigation of surface protons, while reduction of N_2 and H_2O in the synthesis of NH_3, is an unsolved challenge to be addressed. The economy of HER can give clean H_2 stream with reduction of carbon-based fuels as the hydrogen source. At lower temperature and pressure, thermal catalyst can reduce the effect on the environment by the process of ammonia production. Platinum is active for acidic or alkaline media, but comes with higher cost. Chances of improvement in acidic and alkaline media are higher in case of catalyst based on non-precious metal. Its performance is good per unit area for activities of best electrodes having a high capacity of loading. Higher magnitude than non-precious catalyst has effect for both the photocatalytic and electrocatalytic processes. In the first process, incoming light can be obstructed by high loading and in and out the transportation of water and gas can face issues which are of concern be operated in the absence of light. For both media, non-precious HER catalysts development is a challenge [40–45].

5 Grand Challenges

In identifying the optimal ammonia production process that can take place at lower temperature and pressure, there are many challenges that need to be addressed:

i. Discovery of scalable, active, and long-lived catalysts for sustainable NH_3 synthesis. Homogeneous redox/electrochemical/petrochemical surface processes can meet few requirements, but there is no catalyst that can fulfill all requirements.
ii. Enhancement in photochemical and electrochemical approaches for N_2 reduction via electron and proton transfer.
iii. Development of thermal process which can take place at low pressure (<10 atm) and temperature (<200 °C).
iv. Owing the possibility of H_2 production through photo-electrochemical or electrolysis process, it is necessary to find substitute catalyst that can enhance the process, which can operate at lower temperature. Such a catalyst can develop low-pressure process compatible for small-scale production.
v. Enhancement in biochemical routes to N_2 reduction. For sustainable energy into biological process require possible options for enzymes, including nitrogenase, on electrode surfaces.
vi. Development of solar-based thermochemical methods. Nitrogen activation should be enhanced using photons generated by solar thermal process via the metal nitridation/reduction cycles.
vii. Characterization of catalyst structures (chemical, physical, and electronic) and surface adsorbents for favorable ammonia synthesis. Catalyst structures can vary in terms of physical, chemical as well as electronic. The characterization of active catalyst needs to be thoroughly analyzed through numerous experimental studies.
viii Process integration and optimization of catalyst design. This integration provides a breakthrough in catalyst design and provides insights in homogeneous and heterogeneous enzyme catalysis. This approach can be a test bed for more integrated approaches for controlled catalyst synthesis.

6 Critical Steps of the Reaction-Absorption Process

Absorbents at high temperature and presence of ammonia leads to hard solid structure. No impurity should be allowed. The pressure at each stage of the experiment should be checked to confirm proper operation of the recirculation cycle. Recycle of unreacted gases is presented in Fig. 7, which expresses the relationship between pressure change and pump flow. Small reciprocals are in proportion to large chemical reactions. Rate is more at a higher flow of the pump. This limit is near to the fastest reaction, and slope line indicates recycling effects [46–48]. The experiment confirmed that a lower pressure reaction is feasible for ammonia production. In this study, 80% conversion with fast ammonia synthesis rates are obtained. At lower pressure of 25 bars, higher level of production is possible with removal of ammonia. Characteristics of this process are reaction time, separation time, and recycle. The efficiency of the plant is controlled by the chemical reaction rate which is proportional to the time of reaction.

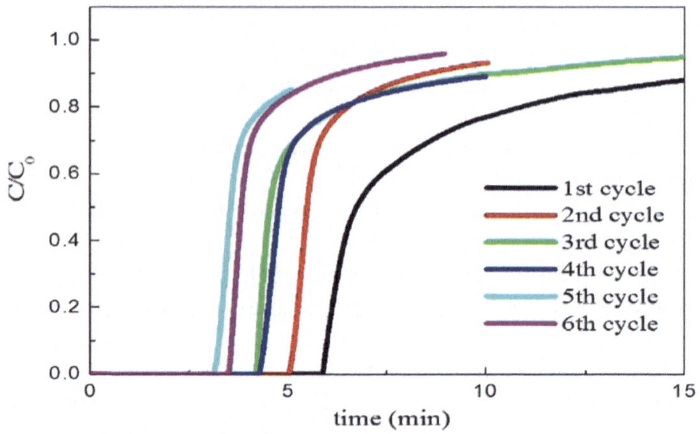

Fig. 7 Performance of absorbents in re-use

7 Conclusions

Low-pressure process is proposed in which N_2 is produced from air using pressure swing adsorption, and H_2 is produced by electrolysis of water, which reacts around 400 °C with catalyst. A thermally catalyzed NH_3 production process is coupled with electrocatalysis or photocatalysis to produce renewable H_2. Solar thermochemical looping is seen as a possible option for sustainable ammonia production, avoiding fossil CO_2 emissions. The studies indicate that the viability of the absorption reaction process for the enhanced production of NH_3 takes place at considerably lower pressures.

References

1. Kletz TA, Amyotte P (2010) Process plants: a handbook for inherently safer design.CRC Press
2. Logadottir A, Rod TH, Nørskov JK, Hammer B, Dahl S, Jacobsen C (2001) The Brønsted–Evans–Polanyi relation and the volcano plot for ammonia synthesis over transition metal catalysts. J Catal 197(2):229–231
3. Galloway JN, Cowling EB (2002) Reactive nitrogen and the world: 200 years of change. AMBIO: A J Human Environ 31(2):64–72
4. Deng J, Iñiguez JA, Liu C (2018) Electrocatalytic nitrogen reduction at low temperature. Joule 2(5):846–856
5. Kandemir T, Schuster ME, Senyshyn A, Behrens M, Schlogl R (2013) The Haber-Bosch process revisited: on the real structure and stability of “ammonia iron” under working conditions. Angew Chem Int Ed Engl 52(48):12723–12726
6. Ertl G (2008) Reactions at surfaces: from atoms to complexity (Nobel Lecture). Angew Chem Int Ed 47(19):3524–3535
7. Chorkendorff I, Niemantsverdriet JW (2003) Concepts of modern catalysis and kinetics, vol 138. Wiley Online Library

8. Sanchez A, Martin M (2018) Optimal renewable production of ammonia from water and air. J Clean Prod 178:325–342
9. Smil V (2002) Nitrogen and food production: proteins for human diets. AMBIO: A J Human Environ 31(2):126–132
10. Tanabe Y, Nishibayashi Y (2013) Developing more sustainable processes for ammonia synthesis. Coord Chem Rev 257(17–18):2551–2564
11. Bielawa H, Hinrichsen O, Birkner A, Muhler M (2001) The ammonia-synthesis catalyst of the next generation: barium-promoted oxide-supported ruthenium. Angew Chem Int Ed 40(6):1061–1063
12. Singh AR, Rohr BA, Schwalbe JA, Cargnello M, Chan K, Jaramillo TF, Chorkendorff I, Nørskov JK (2016) Electrochemical ammonia synthesis. The selectivity challenge. ACS Publications
13. Honkala K, Hellman A, Remediakis I, Logadottir A, Carlsson A, Dahl S, Christensen CH, Nørskov JK (2005) Ammonia synthesis from first-principles calculations. Science 307(5709):555–558
14. Milton RD, Cai R, Abdellaoui S, Leech D, De Lacey AL, Pita M, Minteer SD (2017) Bioelectrochemical Haber-Bosch Process: an ammonia-producing H2/N2 fuel cell. Angew Chem Int Ed 56(10):2680–2683
15. Kozuch S, Shaik S (2008) Kinetic-quantum chemical model for catalytic cycles: the Haber-Bosch process and the effect of reagent concentration. J Phys Chem A 112(26):6032–6041
16. Cherkasov N, Ibhadon A, Fitzpatrick P (2015) A review of the existing and alternative methods for greener nitrogen fixation. Chem Eng Process 90:24–33
17. Damanabi AT, Servatan M, Mazinani S, Olabi AG, Zhang Z (2019) Potential of tri-reforming process and membrane technology for improving ammonia production and CO_2 reduction. Sci Total Environ 664:567–575
18. Dai S-H (1996) A study on the prediction of remaining life and ageing of material for pressurized tubes of industrial furnace operated at elevated temperature. Int J Press Vessels Pip 69(3):247–252
19. Bhaumik S, Rangaraju R, Parameswara M, Bhaskaran T, Venkataswamy M, Raghuram A, Krishnan R (2002) Failure of reformer tube of an ammonia plant. Eng Fail Anal 9(5):553–561
20. Ray AK, Sinha SK, Tiwari YN, Swaminathan J, Das G, Chaudhuri S, Singh R (2003) Analysis of failed reformer tubes. Eng Fail Anal 10(3):351–362
21. Swaminathan J, Guguloth K, Gunjan M, Roy P, Ghosh R (2008) Failure analysis and remaining life assessment of service exposed primary reformer heater tubes. Eng Fail Anal 15(4):311–331
22. Kletz T (2009) What went wrong?: case histories of process plant disasters and how they could have been avoided. Butterworth-Heinemann
23. Holland M, De Bruyn H (1996) Metal dusting failures in methane reforming plant. Int J Press Vessels Pip 66(1–3):125–133
24. Bharadwaj S, Schmidt L (1995) Catalytic partial oxidation of natural gas to syngas. Fuel Process Technol 42(2–3):109–127
25. Singh J, Basu P, Rao B (2002) Fire in secondary reformer outlet line to waste heat boiler. Process Saf Prog 21(3):205–211
26. Jahromi S, AliPour M, Beirami A (2003) Failure analysis of 101-C ammonia plant heat exchanger. Eng Fail Anal 10(4):405–421
27. Kunjunny A, Patel M, Nath N (1999) Revamping of CO ~ 2 removal section in ammonia plant at IFFCO Kalol. Fertiliser News 44:53–58
28. Harjac S, Atrens A, Moss C (2008) Six Sigma review of root causes of corrosion incidents in hot potassium carbonate acid gas removal plant. Eng Fail Anal 15(5):480–496
29. Pande JO, Tonheim J (2001) Ammonia plant NII: Explosion of hydrogen in a pipeline for CO2. Process Saf Prog 20(1):37–39
30. Alhabdan F, Elnashaie S (1995) Simulation of an ammonia plant accident using rigorous heterogeneous models: Effect of shift converters disturbances on the methanator. Math Comput Model 21(4):85–106

31. Krivshich N, Pavlyuk S, Deineka A, Kolesnik S (2003) Introduction and use of dry gas seal systems on ammonia plant compressors. Chem Pet Eng 39(9):608–611
32. Anwar Z, Ahmed S, Kirmani IA (1997) Expansion bellows failure of synthesis loop hot heat exchanger. Process Saf Prog 16(2):105–109
33. Shah M, Zhu G (1998) Burst resistant ribbon wound pressure vessels for ammonia plants. Process Saf Prog 17(2):98–103
34. HOLNESS DL, PURDHAM JT, NETHERCOTT JR (1989) Acute and chronic respiratory effects of occupational exposure to ammonia. Am Ind Hyg Assoc J 50(12):646–650
35. Khan MR, Kabir AZ (1995) Availability simulation of an ammonia plant. Reliab Eng Syst Saf 48(3):217–227
36. Walter MG, Warren EL, McKone JR, Boettcher SW, Mi Q, Santori EA, Lewis NS (2010) Solar water splitting cells. Chem Rev 110(11):6446–6473
37. McCrory CC, Jung S, Peters JC, Jaramillo TF (2013) Benchmarking heterogeneous electrocatalysts for the oxygen evolution reaction. J Am Chem Soc 135(45):16977–16987
38. Suntivich J, May KJ, Gasteiger HA, Goodenough JB, Shao-Horn Y (2011) A perovskite oxide optimized for oxygen evolution catalysis from molecular orbital principles. Science 334(6061):1383–1385
39. Callejas JF, McEnaney JM, Read CG, Crompton JC, Biacchi AJ, Popczun EJ, Gordon TR, Lewis NS, Schaak RE (2014) Electrocatalytic and photocatalytic hydrogen production from acidic and neutral-pH aqueous solutions using iron phosphide nanoparticles. ACS Nano 8(11):11101–11107
40. McKone JR, Sadtler BF, Werlang CA, Lewis NS, Gray HB (2013) Ni–Mo nanopowders for efficient electrochemical hydrogen evolution. ACS Catal 3(2):166–169
41. McEnaney JM, Soucy TL, Hodges JM, Callejas JF, Mondschein JS, Schaak RE (2016) Colloidally-synthesized cobalt molybdenum nanoparticles as active and stable electrocatalysts for the hydrogen evolution reaction under alkaline conditions. J Mater Chem A 4(8):3077–3081
42. McCrory CCL, Jung S, Ferrer IM, Chatman SM, Peters JC, Jaramillo TF (2015) Benchmarking hydrogen evolving reaction and oxygen evolving reaction electrocatalysts for solar water splitting devices. J Am Chem Soc 137(13):4347–4357
43. Lu Q, Hutchings GS, Yu W, Zhou Y, Forest RV, Tao R, Rosen J, Yonemoto BT, Cao Z, Zheng H, Xiao JQ, Jiao F, Chen JG (2015) Highly porous non-precious bimetallic electrocatalysts for efficient hydrogen evolution. Nat Commun 6(1)
44. Kibsgaard J, Chen Z, Reinecke BN, Jaramillo TF (2012) Engineering the surface structure of MoS2 to preferentially expose active edge sites for electrocatalysis. Nat Mater 11(11):963–969
45. Hsu C-L, Chang Y-H, Chen T-Y, Tseng C-C, Wei K-H, Li L-J (2014) Enhancing the electrocatalytic water splitting efficiency for amorphous MoSx. Int J Hydrogen Energy 39(10):4788–4793
46. Jung J, Jeong YS, Lim Y, Lee CS, Han C (2013) Advanced CO_2 capture process using MEA scrubbing: Configuration of a split flow and phase separation heat exchanger. Energy Procedia 37:1778–1784
47. Dave N, Do T, Puxty G, Rowland R, Feron PHM, Attalla MI (2009) CO_2 capture by aqueous amines and aqueous ammonia–a comparison. Energy Procedia 1(1):949–954
48. Ciferno JP, DiPietro P, Tarka T (2005) An economic scoping study for CO_2 capture using aqueous ammonia. Final report, vol 28